Advanced Wreck Diving Guide

Tom Packer poses for a shot with the *Andrea Doria*'s bell. With some of the anemones scraped off, the true bell shape is visible.

Advanced Wreck Diving Guide

by Gary Gentile

Cornell Maritime Press, Centreville, Maryland

Library of Congress Cataloging in Publication Data

Gentile, Gary, 1946-
Advanced wreck diving guide.

1. Diving, Submarine. 2. Shipwrecks. 3. Underwater archaeology. 4. Excavations (Archaeology) I. Title.
VM981.G46 1988 627′.72 87-47997
ISBN 0-87033-380-1

Manufactured in the United States of America
First edition, 1988; fourth printing, 1994

Contents

Introduction *vii*

1. Dive Boat Safety 3
2. Overnight Dive Trips 14
3. Wreck Penetration 25
4. Deep Diving Procedures 39
5. Decompression Methods 49
6. Artifact Recovery 72
7. Artifact Preservation and Restoration 89
8. Quick Photography Techniques 108
9. Shipwreck Research 118
10. Lobster: Tail or Tale 127

Introduction

Wreck diving has come of age.

Ever since man first saw the sea, he has been enthralled by it. He swam in it; he built boats and ships to sail upon it; he constructed great ports from which to run foreign commerce. Eventually, his whole civilization revolved around the use of the sea. The oceans enabled man to migrate to the far corners of the world. The great biomass of marine life fed him. The endless reaches of the waves gave him cause to wonder.

Yet, in all these thousands of years of prehistory, he had no knowledge of what was below the thin layer of surface tension other than what he could see in the shallow water, or what he pulled up in his fishing nets. The ocean floor was an alien environment which his primitive imagination peopled with strange beings and mythological creatures.

Man's attempt to see what lies beneath the surface by breathing compressed air supposedly goes all the way back to the fourth century B.C., when Alexander the Great descended in an inverted bucket to the bottom of the Mediterranean. It was an undertaking only a king could envision. Today, anyone with the desire and the initiative can dive below the bounding main to explore the unknown and unclaimed depths.

The longing to visit the underwater realm is deeply imbedded within us all. Satisfaction of this longing, which started in man's archaeological past, is now within the grasp of almost everyone. The limitations of diving are fast disappearing—due to reliable and easy-to-use equipment, advanced technology, and nationally recognized training programs. Still, self-contained diving is only in its beginning stages.

Whereas in mountain climbing there is always another peak, in diving there is always another trough—and in wreck diving, always another wreck. Wreck diving has become a category of its own. With it, as with any specialty, has evolved a distinct brand of expertise, as different from commercial diving as racing a sports car in the Indy 500 is different from steering an eighteen-wheeler. We may all be divers, but shipwrecks offer unique conditions which have led to specific and highly esoteric disciplines.

Wreck diving is more than just a vogue. It is a form of delving into the past, of re-creation of history, of exploration of man's nautical beginnings. In many cases, sunken vessels are the only extant examples of the bygone days before the mast. Rotting, or rusting, hulls recall the glory of sail, perpetuate the era of the excursion steamer, or bring to mind the awful tragedy of war.

All this and more is to be found in the vast repository of the sea. To the wreck diver all doors, and hatches, are open. You need only look.

Advanced Wreck Diving Guide

A diver hits the water.

CHAPTER 1

Dive Boat Safety

Divers, whether they like it or not, in accepting a certification card, are consigning themselves to the sea.

There is a limited amount of non-ocean diving, such as diving in caves, quarries, rivers, and inland lakes or ponds. And there are divers who submerge their wet-suited bodies in golf course traps for their dimpled rewards. But by and large, in the words of the vernacular, "the ocean's where it's at."

Whether the attraction is wrecks, reefs, or spectacular marine organisms, ocean diving has become a way of life. But where the diving courses end is where the real diving begins. There is an awful lot one needs to know, and observe, in order to make the topside part of the sport as safe as the underwater environs.

Dive boats come in many sizes and shapes, from the sleek converted crew boats of the Gulf, to the "head" boats of the

Eastern Shore, the luxurious, diver-oriented, and ergonomic West Coast yachts, and the Caribbean types: hundred-foot sailboats with crews and overnight accommodations, and island-hoppers constructed of an outboard motor lashed to a raft of boards and fifty-five-gallon drums, with a canvas shade. They come with or without guides, with or without captains, and with or without licenses. Let's examine these points one by one.

Unless you rent or own your own boat, the license is the single most important item you should look for. In the United States, the Coast Guard has strict rules and regulations with which every charter boat must comply in order to be acceptable as a passenger-carrying vessel. These safety regulations require such things as a fire-retardant hull, watertight bulkheads, fire-fighting equipment, communication devices, and adequate lifeboats and life preservers for all aboard. In addition to monitoring a boat during its construction, the Coast Guard inspects each and every vessel annually, ensuring that safety gear does not fall into disrepair and that new ordinances meet with compliance.

The Coast Guard requires that the license be prominently displayed, and on it you may read the limits the boat is allowed. For instance, for a vessel licensed for more than six passengers to be permitted to go more than twenty miles offshore, the boat must be equipped with a single-sideband radio and an EPIRB (electronic position-indicating radio beacon), a device that is activated by contact with salt water, and thereupon emits a continuous radio signal so the Coast Guard is alerted automatically and rescue craft know where to search. It's impossible to go into the thousands of Coast Guard regulations here, but at least be aware that you do not want to be seventy miles at sea with a dozen companions, in a boat licensed for six at twenty miles: you may be the one without the life preserver should an accident occur. Also know that you have legal grounds for walking off a boat without paying should you arrive at the dock and find that the captain has made arrangements for which his boat is not licensed.

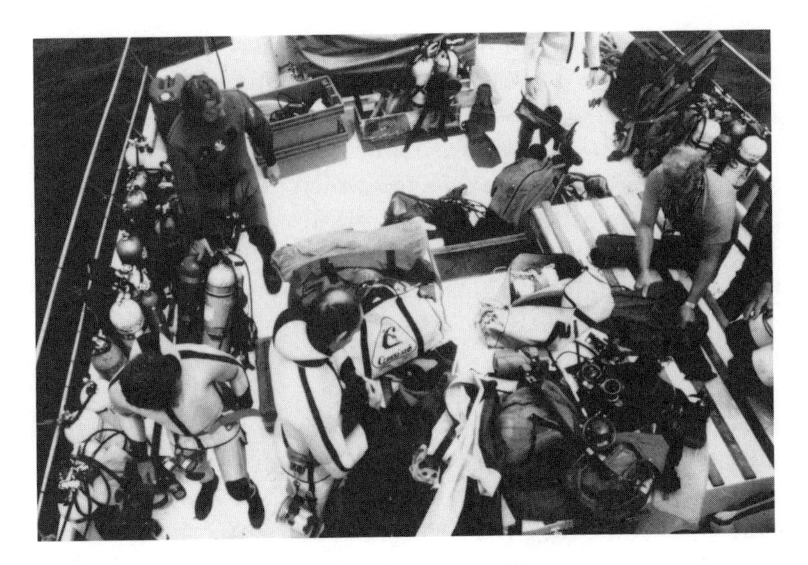

Divers getting dressed.

Captains, too, require licenses, but in this respect the Coast Guard, although it tries, is not able to maintain standards as stringent as those for boats. To qualify for a captain's license, a person needs 365 days of sea time, as a working mate, in addition to a passing grade on a test which includes such nautical knowledge as Rules of the Road, navigation, and safety. The test is not easy, and the Coast Guard thinks nothing of flunking someone who gets less than a perfect score in Rules of the Road. However, the apprenticeship time can easily be fudged. A person taking the test must have proof that he has the required time at sea. This is obtained by getting captains for whom he has worked to "sign his time." If you are friendly with three or four captains you can get each to sign for ninety days, and this is usually done as a courtesy. Or, if you have your own boat, even a sixteen-footer, you can sign some of your own time. Since most dive boat captains reach that position after first having been divers, this part of the requirement is a joke. I know of only one captain who has actually served all the days as a working mate that were required.

On the other hand, perhaps the Coast Guard realizes the ease of fudging apprenticeship time and has compensated for it in the time allowance. In any case, the best way to choose a captain is by reputation. Advertisement in a magazine does not make a reputation; word of mouth does. In this respect, your fellow divers are the ones who can speak with experience on which captains are good and which captains are bad. Do not get confused into believing that a good captain, who is capable of handling his boat in a seamanlike manner, is necessarily a good person. Too many captains are too influenced by greed and mortgage payments to be able to make good judgments relating to safety and foul weather. I have walked off boats when captains said that, even though the Coast Guard was flying small craft advisories (two flags, for gale warnings), they were going to sail, just so they could collect their charter fee. Unfortunately, the Coast Guard does not have any regulations in this matter, only "advisories." They leave judgment up to the captain, who is, by nature of his license, competent to make such judgments. Unfortunately, a captain may

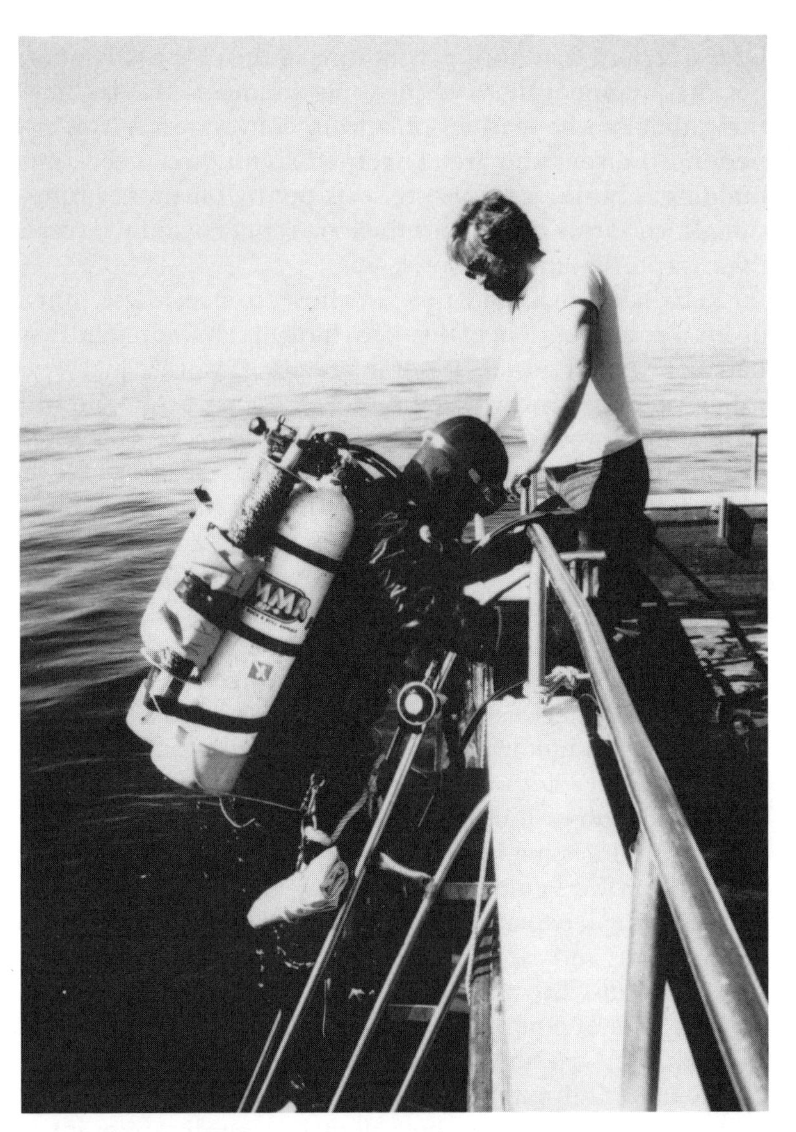

The mate standing by in case the diver needs help with his gear as he climbs up the ladder. Note placement of the decompression reel, between right tank and pony bottle.

be biased and therefore may not make the safest judgment, but the judgment that benefits him the most. This is something that can be learned only from conversations with experienced divers who are concerned about diver safety, not building a business. It also reflects poorly on those professionals who truly care about their passengers, and who take their responsibilities seriously.

Let's backtrack, and find out how to meet these other divers whose experience you might have to rely upon. The first person to ask for advice is your instructor. If you are lucky, he will be an experienced diver as well as an instructor, and will be able to tell you who to contact in order to get on boat dives to your desired locations. If not, he should at least be able to put you in touch with other divers who can supply the information.

Many instructors work directly for dive shops that run their own charters, either as a part of the business or as a courtesy to keep the students diving (and buying equipment). If they work for a certifying agency or on a free-lance basis, they most likely have the necessary contacts. If they work for a dive club, they will undoubtedly try to steer you in the direction of joining the club. Either of these choices is a good one, for dive shops and clubs will have charters already set up, a schedule of dates and destinations, and qualified dive masters. Thus, they will provide guidance on signing up for dives that suit your needs and experience level, and you can rely on them to choose boats and captains.

Inform the dive master about your interests, experience, and the type of equipment you own. If he is a good DM, he will talk with you about what to expect on the dive, find a buddy for you if one is required, offer advice and cautions, and give encouragement that will make your dive pleasant as well as safe. At the same time, he will also expect you to participate in dive boat etiquette, and even to help others should the need arise. Never have the attitude that because you are paying your freight you can simply sit back and relax while others do all the work.

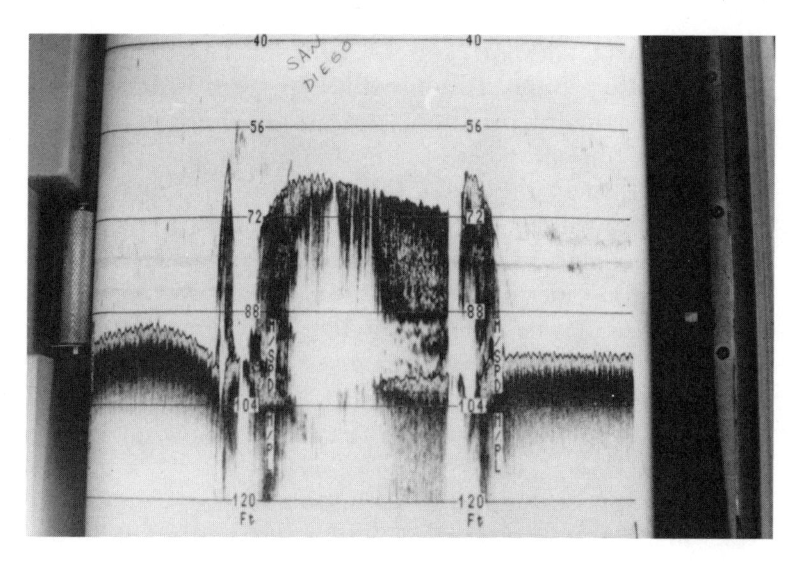

A good depth recorder is essential in locating shipwrecks.

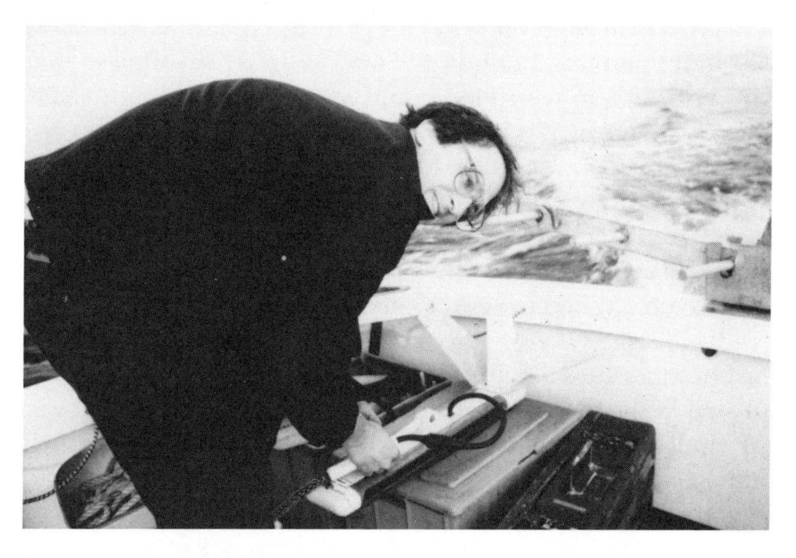

Side scanning sonar is expensive, but much more sensitive and farther reaching than the electronics carried on the average dive boat. Here the towfish is prepared for dragging.

Diving is a team sport, and it requires that each individual make an effort to contribute whatever he can. It is not a labor/management situation. You are all there to aid one another. If someone needs help, whether on board (pulling on a glove), or in the water (getting back to the boat), be prepared and willing to do your part. If you're going to say, "It's his job—let him swim out a line," then you should quit diving before your inaction causes a fellow diver grief. Remember, you may be the one needing help some day. Divers are all doers, and leadership can be assumed by anyone, depending on the circumstances. If you are the closest and the ablest, then lend a hand. Don't wait for someone else to do it.

Let us suppose that you are diving as a "walk-on." In a resort area you can simply call a dive charter service, find out when they have a head boat leaving the dock, show up, and dive. You are a total stranger to the operators as they are to you. Check in with whoever is in charge (the captain, dive master, or guide) and find out how the operation is run—first so that you know it's to your liking, and second so you don't cause any problems by not conforming to their rules. Chances are they have been in business long enough to know the hazards, and probably, because of the great number of unknown divers they encounter, they have strict methods of keeping an eye on their customers. Believe me, they get a lot of "turkeys," so they're only trying to protect themselves.

In this case, diving with a guide might be the way to go—not because he's a better diver than you, but because he can probably show you more than you would see on your own. After all, he knows the territory. If you're taking pictures, tell him what you want to photograph and ask if he thinks he can accommodate you.

If you don't want to be led by the hand, ask around to see which other divers have similar interests and experience. But don't fall into the mistake of choosing a more experienced diver in order to make a dive you otherwise shouldn't be making. This could be disastrous for you should you become separated. And be very cautious when choosing dive partners

A Coast Guard helicopter lowers a basket for a stricken diver.

from among strangers. Because you don't know them, you shouldn't rely on them. After all, they may be pure bluff, inflating a needy ego at your expense. If you really don't like or trust anyone on the boat, and there is no guide to dive with, you're better off diving alone. You, at least, know your own abilities and limitations, and should exercise enough caution not to exceed them.

Never be afraid to take things into your own hands. Take a moment to walk around the boat and learn the layout: where the safety equipment is stored, where the water entry is to be made, and especially where the ladder is located. Mingle a bit and learn what you can about the reef or wreck site. Make friends. One of the most rewarding aspects of diving is being with people who have similar interests.

You can prepare yourself for a dive by checking out land-based facilities: the nearest recompression chamber, and transportation to it. A call to the Divers Alert Network will usually tell you which hospitals are operating chambers at the time. In the United States, the Coast Guard is the most valuable organization you will encounter. They keep a constant vigil on specified radio frequencies, have boats ready at the dock and on patrol, and often have helicopters that can pick up stricken divers.

Other countries do not have these arrangements, so you will have to find out what safety precautions are available by asking the dive charter service. Don't let them pooh-pooh you with evasive answers. Remember, it's your life they're pooh-poohing. Always take the initiative.

Once, when I was dive mastering, the wind kept increasing during the course of the day until, by afternoon, while we were still anchored into the wreck, the ocean was quite lumpy and a strong current was running. Divers were going into the water, but having a rough time of it. Standing safety on the bow, I held a life buoy attached to a rope, ready to toss it to anyone unable to combat the sea conditions. One diver stood beside me, observing those in the water. Finally, he said, "I don't think this is for me," and started getting undressed.

I respect him, and his opinion. Some were willing to fight the current and waves in order to make that last dive. He made up his own mind, knowing how he felt on that particular day, and not being intimidated by the false bravado of his comrades. There is nothing wrong with being concerned about your own safety. Anything you consider unsafe, you don't have to participate in. Certainly, because of differing experience levels, what may be considered a risk for you may not be a risk for someone else. Don't judge others by your ability, and don't let others judge you by theirs. You are the one with the final say about what you do. Ultimately, it is your decision that counts the most.

CHAPTER 2

Overnight Dive Trips

Diving has come a long way in recent years, with millions of certified divers now plying the world's oceans in search of adventure, excitement, relaxation, wonder, and awe. Where once people were content to slip into the waves from the safety of the beach, or from a rock jetty, they now spend the entire day on a boat in order to reach seldom-visited offshore sites.

As a logical extension, combining efficiency with the week-long "standard" vacation, more and more people are seeking out dive trips that extend to several days, even a week. No longer bound by an eight-hour day, divers can travel farther, make more dives, optimize bottom time, have longer surface interval, and enjoy the camaraderie of fellow divers with more intimacy. Because the need for haste is eliminated, they are not rushed to get in and out of the water according to the captain's time schedule, and the diving is safer as well as more restful.

Anchored in one spot, reef divers can observe the marine life at different times: morning, afternoon, and night. Wreck divers can move around and see many wrecks, or study one in detail. There is sufficient time to photograph the local environment with a variety of lenses, from macro to wide-angle, or to fan the sand for precious artifacts and dig deep before the ever-moving sea fills in the hole. In addition, the diver who stays in one area for an extended time will not have to waste valuable bottom time reorienting himself on each dive.

Multiple-day dive trips have much to offer. But, as in any new undertaking, there are things you need to know. The diver who is forewarned and forearmed is foremost.

Safety

Many Caribbean operations are running dive cruises, either on large sailing ships or on charter boats offering fine appurtenances. These vessels are usually fully crewed, and offer meals and air-conditioned staterooms. There is little the diver need think about other than diving and having a good time. But as the local dive boats get into the overnight business, they are not always so well equipped. Often, because of space limitations and cost-effectiveness, the captain relies on the charter to help out. In any case, the diver should know some of the basic rules and hazards of nighttime anchoring in the open ocean.

According to Coast Guard regulations, a captain may work only twelve hours. Technically, if the boat stays out any longer, it needs a second captain; usually, an experienced mate will suffice. Once, I was on a dive charter where the captain handled everything himself and relied on the divers to help him drop and pull up the grapnel, a job we were willing to do. After a six-hour ride to the wreck site and a hard day's diving, I was too exhausted to wonder how the captain managed to stay awake during the return to port. It was not until we docked that I learned he simply put the boat on autopilot and climbed into his bunk for a few hours of sack time!

On this kind of trip, where you are "chipping in," you should know something about small-boat handling, and be aware of important details such as the placement of fire extinguishers, location of life jackets, and basic emergency procedures. If the captain also dives, those on board should know how to start the engine, use the radio, and call the Coast Guard.

Sharing the workload is accepted practice. Divers can take turns at the helm and stand watches at night. As military as it might seem, someone must always be alert. Several years ago some friends of mine were sleeping soundly when they were rudely thrown out of their bunks. Amid splintered wood and shattered glass, they clambered topside in time to see the immense steel hull of a freighter scrape along the side of their thirty-foot boat. Fortunately, they had not been rammed head-on, or their story would never have been told. Still, the flying bridge was knocked off and lost, and the rail and gunwale smashed. They limped home in sinking condition, unable to broadcast for help. The freighter did not stop or make a report: those on board were probably as unaware of the collision as you are when your car runs over an insect.

Radar is a must. In fog or in the dark, you must have constant awareness of your surroundings. If blips on the screen appear to be too close, wake up the captain. And keep a knife or axe on the bow in case you must cut your anchor line in a hurry. A marker buoy previously tied to the anchor line, and large enough to keep the rope afloat, will allow you to return to your mooring without difficulty—unless the ship has carried it away.

What to Bring

If you think that in addition to your normal diving gear you need only bring a toothbrush, a tube of toothpaste, and a lasting smile, you are sadly mistaken.

Peter Hess displays the latest fashion in underwater head gear.

Unless the charter boat supplies linen, the first thing you will need on an overnight trip is a sleeping bag. Even in summer the ocean air becomes chilly after sunset. By the same token, extra clothing should include long pants, long-sleeved flannel shirt, sweater, and windbreaker. A visor or peaked hat, in addition to sunglasses (which are breakable), can save your eyes from hours of anguish. Foul-weather gear is a must, as are galoshes. And leave those flip-flops at home: the decks are too smooth, the seas are too rocky, and there are too many hard objects to bump your toes on. Wear boat shoes or sneakers for protection as well as for grip. And when you're not wearing clothes (other than a bathing suit), keep a goodly supply of sun block on hand. Two or three towels can be rotated to keep you dry.

Dampness is part of boating, and salt spray will find its way into every corner of the boat, including the insides of your bags and suitcases. Air conditioning helps reduce interior humidity, but even so by the end of a week your sleeping bag will have gained some weight, balled-up t-shirts will grow mold, and all your clothing will be impregnated with salt. You will begin to feel—icky. The way to slow down this process is to pack your things individually in plastic bags. Everything, including your sleeping bag, must be washable.

Pizza shops do not make offshore deliveries, so unless the boat supplies food you have to bring your own. Learn in advance what facilities are available. If there is no refrigerator or freezer, you must pack everything in coolers—and even block ice lasts only two or three days. Be careful of frozen carbon dioxide (dry ice), because the intensity of the cold will crack the plastic lining of most coolers: insulate it with a towel. Boat coolers should have a raised lip on which the lid closes; without this, if the cooler is kept out on the deck, salt spray rolling off the top will work its way inside, and you will find your food floating in a puddle of seawater.

If you do not put into port for refueling, you will have to plan to use canned rations; or, if there are cooking facilities on board, take freeze-dried backpacking meals. Carry plenty of your favorite snacks, such as cookies, potato chips, and pretzels. Nibbling is a tasty way to offset salt loss from perspiration.

Cold cuts and crackers get old after a while, so nowadays many boats carry a generator and a microwave oven. With the wide assortment of dinner entrees available, you can have a varied and delectable diet. It is but the work of a moment to make hot soup or heat up water for instant coffee. You can even "nuke" your sandwiches and have hot ham and melted cheese. And don't overlook microwaved popcorn.

Perhaps more important than food is drink. It is easy to become dehydrated in the heat, sun, and wind, especially when the water aboard tastes like the inside of a rubber tire. Additives such as Kool-Aid, Crystal Light, iced-tea mix, and bouillon can make all the difference in the world when space for juices and bottled soda is limited.

Paper cups, bowls, and plates, and plenty of trash bags, will remove much of the burden of cleaning up on a crowded dive boat. Bring old silverware in case it is lost. Because of sloppy seas, a narrow-necked bottle you can wrap your lips around, or a plastic mug with a lid and sip hole, is essential for getting liquids down your throat instead of on your neck.

Finally, you can always catch lobster or fish for true seaside cuisine.

Backups

There are no dive shops or hardware stores at sea. If your equipment breaks, you must either repair it yourself or use a spare.

Since diving is a methodical sport that requires certain organizational skills, most people will already have a number of parts and extras around the house. I suggest everything be gathered into a tackle box with compartmented trays and always carried along.

People usually think of things like extra mask straps, fin straps, and O-rings, but suppose your tank blows a rupture disk? Dive boats with compressors are not likely to have water-filled troughs, so tanks overheat quickly during refill or while sitting in the hot sun on the open deck.

Take a junk box and fill it with nuts and bolts, rubber bands, paper clips, tie-wraps, hose clamps, needle and thread, string and nylon lines, electrical tape, and chewing gum. You would be surprised how many repairs can be jury-rigged. A tube of neoprene cement and a hair dryer are essential for fixing wet suits and dry suits. Carry a patch kit for your BC (buoyancy compensator) and liftbag, too.

A spare high-pressure hose for your tank pressure gauge is important. When traveling out of the country, or to out-of-the-way places, you might consider taking along a rebuilding kit for your regulator. Even a person skilled in regulator service or repair is lost without the parts. A complete spare regulator is not a bad idea, either.

If your lights or strobes use nickel-cadmium batteries, make sure the boat has the appropriate voltage for recharging. Otherwise, carry spare battery packs, or use equipment that relies on alkaline cells, and have plenty. Don't forget spare bulbs.

Consider the fate of a fellow diver who made a spectacle of himself on the highway. His glasses were smashed on the dive boat, and he was forced to wear his prescription mask home. He got same strange stares at the toll booths. Even those using extended wear contact lenses should carry a spare pair of glasses in case of loss or breakage. Don't forget a container in which to keep the contacts moist.

In addition to your favorite brand of motion sickness preventative, remember to bring any prescription drugs you are taking, or medicine for ailments to which you are prone. Aspirin or acetaminophen will relieve minor headaches and muscle pains attributed to the strain of boat diving, and always keep on hand sinus medication to facilitate ear clearing during descent, and to break up post-dive reverse blockage. If you think the days of house calls are over, try to get a doctor to make a boat call.

Take along a tool kit with the essentials: standard and Phillips screwdrivers, adjustable wrench, pliers (slip-joint, pump

A curious onlooker, a bottlenose whale surfaces right next to the boat.

A sand tiger shark in a school of baitfish.

type, and wire cutters), and O-ring extractor. If any of your equipment requires specific tools, such as fine screwdrivers, spanner wrenches, or allen wrenches, make sure you have them. You are the ultimate repairman, and need to be self-reliant.

Then there are underwater tools: hammers, chisels, and crowbars. A small can of silicone can keep them from rusting too badly if you spray after each immersion; your zippers will love it, too. And a couple of extra liftbags never hurt.

If you're already exhausted from gathering all this paraphernalia, ponder the plight of the underwater photographer. Cameras and strobes are too complicated to be repaired by anyone but expert technicians. Checks can be run with an inexpensive voltage-tester/ohmmeter, and a few minor adjustments can be made. Beyond that, you're in a bind.

The only solution is a complete system backup, with interchangeable parts: if the camera malfunctions, fit another to the tray; if the strobe won't recycle, replace it with a spare. If the trouble is in the cord or connector, substitute a new one. Routine flushing with fresh water will prevent salt encrustation and electrolytic corrosion. Most important of all, tear down your camera system completely at the end of every diving day and rinse off all threads; otherwise, bolts and synch cords will became permanently bonded to their sockets. An old toothbrush is essential for removing sand particles and hardened O-ring grease. And for the real paranoiac, I suggest alternating your camera usage, just in case one has an internal malfunction which is not obvious, but which prevents images from being passed onto the film.

Expensive? Yes. But consider the frustration of spending hundreds, or thousands, of dollars for a week-long trip, possibly incurring wage loss, and then having your camera stop working the first day. If you are a serious photographer, and photography is your primary aim, then you must view the overall financial picture. Consider the extra equipment not as a loss, but as an investment. And don't forget to bring plenty—*plenty*—of film.

Pastimes

Very little of your time on a multiple-day dive trip is actually spent underwater. Mostly you hang around on the boat trying to find a comfortable place to sit. Some people can prattle on for hours, but I soon tire of hearing old divers' tales. I like to retire to a bunk and catch up on my reading, or push graphite across foolscap. Others play cards, smoke, and swear, as if the boat were a Mississippi riverboat loaded with gamblers and gunslingers.

Keeping your gear in order, and making sure it does not get mixed up with that of your fellow divers (even if it is labeled), is a full-time job. However, there are other options available to you.

If the captain does not already have fishing gear, you might consider bringing along rod and reel and lots of lures. Besides enjoying the sport of catching a big one, you can cook up the fresh fish.

During the day you can observe pelagic marine life: everything from plankton to whales will eventually swim past an anchored boat. At night, away from the lights of civilization, the stars come out in full panoply, often visible as soon as they peek over the horizon. With a simple star guide you can pick out constellations you might never see at home. And the eventide bioluminescence can be quite a treat.

I was on one boat where the captain brought a television and a videotape player. Let me tell you, rolling with the stern Atlantic swells while watching the storm sequence of *Das Boot* is a wild experience!

Caution

I must close on a serious note. While your mind and heart are losing their inhibitions with newfound friends, your body is quietly building up residual nitrogen. Those using one of the new decompression computers on the market need not worry, as long as they keep their units switched on. But if you rely on the *U.S. Navy Standard Air Decompression Table,* you must throw in a fudge factor.

Recent studies have proven that some slow tissues do not release all the absorbed nitrogen within the twelve-hour interval specified by the *Tables.* After several days of diving—even nondecompression diving—you may accumulate a dangerous concentration of the gas. I recommend strongly that on multiple-day dive trips all dives after the first, unless a day is missed, be considered as repetitive dives. Even after the twelve hours prescribed by the *Tables,* for total desaturation, safeguard yourself by using the next deepest depth for the next time increment.

You may never know if this precaution was necessary. But then, you'll never be wrong.

CHAPTER 3

Wreck Penetration

Wreck divers are not born, they evolve.

From swimming pool to quarry to open ocean, each diver must find his niche, his prime area of interest. For some it will be beautiful coral reefs and exotic marine life; for others it will be shipwrecks. For those who feel the pulse of excitement in exploring man's sunken history, in bringing back mementoes of the past and of their experiences, the ultimate challenge is the penetration of these collapsing and encrusted ships: exploring twisted corridors, entering hidden rooms, throwing light on the all-encompassing darkness within. This is both a thrill and an adventure. And it is completely different in approach, with its own methods, skills, and stringent safety rules.

Backup is the prime word in wreck penetration. In addition to the usual practice of buddy diving, equipment backups are vitally important. As soon as a diver enters a wreck, he gives up his most important means of escape—the free ascent. Without the option of screaming for the surface, he must have alternative moves for every contingency: running out of air, regulator failure, entanglement, becoming lost or disoriented.

This breaks down into two categories: equipment and planning.

Wrecks close to shore, or in shallow water, are highly prone to the devastating actions of the sea: waves, storms, currents, oxidation, and destructive marine organisms. As hazards to navigation they are often demolished by explosives or wire dragging. For that reason, most intact shipwrecks are in deeper water. This means you need a large air supply.

Double tanks are a must. If they are unyoked, with a regulator and submersible pressure gauge on each tank, the system also offers a backup in case of regulator failure. However, it necessitates a switchover during the dive from one regulator to another. If you haven't planned properly you might find yourself emptying one tank completely, thus obviating its use as a backup, and you might have to change mouthpieces at a time when you are otherwise engaged. It is best not to have to stop in mid-dive if it can be avoided.

The use of a yoke between tanks permits one regulator to be used for the entire dive. Some people like to use an octopus rig: one first stage with two second stages. The extra second stage can be easily passed to a buddy, or slipped into the mouth should the main second stage fail. But this offers no backup in case of first-stage failure.

The best system is to add a pony bottle: a small, extra tank clamped to the rear of the doubles, containing an additional twenty or so cubic feet of air. With its own regulator it becomes a complete, separate breathing source.

This second regulator must be readily accessible. It does no good hanging to one side where it can catch in rusting I beams, or dangling behind out of reach. Usually, if you need to go to your backup, you cannot take the time to study a road map in order to locate it. You must be able to put your hand on it instantly. A lanyard clipped to a tank harness or BC strap

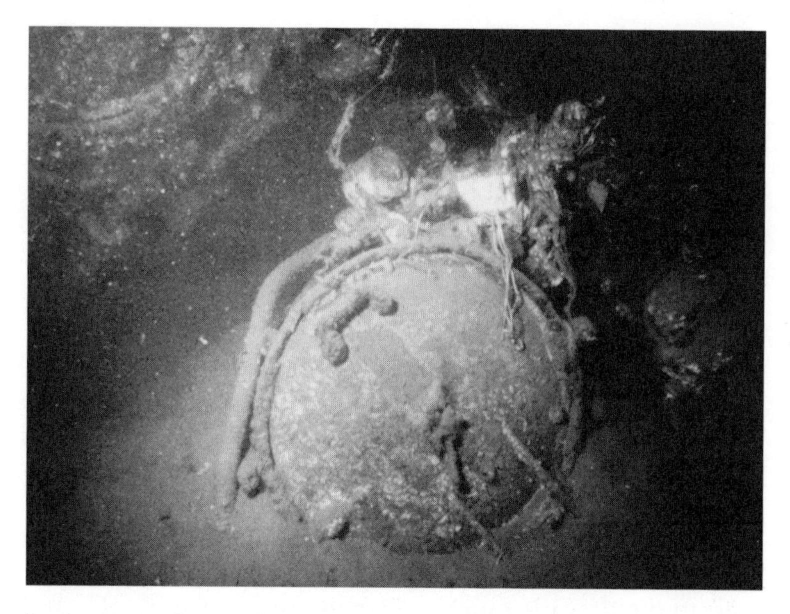

In the forward torpedo room of the *U-352* a torpedo tube lies partially buried in the silt.

keeps it in front of you, as does a neck loop around the mouthpiece. I always remove the exhaust tee so the second stage stays flat against my chest.

It is also important to differentiate one second stage from another, so you will not accidentally drain your pony bottle. Using regulators of different manufacture, or mouthpieces of different types, or placing a colored tape on the hose next to the second stage, are ways of doing this. Also, you can exchange equipment checks with your buddy on the anchor line before making your descent. Remember, if he gets into trouble, you get into trouble.

The next most important piece of equipment is not just a light—but a spare light. The most powerful light you can afford is what you should use as a primary. Lights with rechargeable nickel-cadmium batteries offer the brightest beam, but they last only an hour to an hour and a half, and when they start to die they die quickly. They should be freshly charged before each dive.

But the backup cannot be a low-power, low-quality light. It must be bright and reliable because if your main light source fails through battery death, bulb filament breakage, or flooding, the backup becomes your main, and you must now find your way out of the wreck with it. It does not need to be as bright as the primary, but bright enough to do the job. Using a backup with alkaline batteries ensures long life, because these will die very gradually, not suddenly.

The spare light needs to be accessible—not tucked away at the bottom of your goody bag under other gear. Carrying it on your arm, with a wrist lanyard, is the best way. The spare light must also be tested immediately prior to penetration. When you need it the most is not the time to find out it does not work. It is even a good idea to have a second light in the "on" position, and use a third as a spare. The first time your light goes out inside a wreck, and you experience the fear of absolute blackness, you will agree.

Because of the possible presence of interior snags, cables, and lines from previous divers, two knives are recommended:

Inside the conning tower of the *Tarpon.* Note the gyrocompass repeater above the telegraph.

one on your leg, and one on your gauge console in case reaching your leg should prove awkward.

Since wrecks seldom offer spacious accommodations, you are often crawling through them instead of swimming freely. This means the knees get an inordinate amount of wear, especially if you pray a lot. Knee pads help protect dry suits from scrapes and puncture wounds, and I've developed my own design, called "garypads." These are made from a car inner tube sliced into twelve-inch sections. The back of the knee is cut out, leaving a one- or two-inch circular band at top and bottom. This way the pads can be pulled on easily and will allow the leg to bend without resistance. They are amazingly tough, yet flexible enough to provide adequate movement for both swimming and climbing up the boat ladder. Make sure you get the right size for your leg: some divers will need tubes from Volkswagen tires, others from Continentals. You can usually get old ones for nothing from your local auto mechanic.

Now for some things not to carry into a wreck: large goody bags, heavy tools, liftbags, decompression reels. The penetration diver needs to be streamlined, without any projecting gear that can snag on wreckage. All items unnecessary for penetration should be left outside the wreck, to be retrieved later. If you want to carry a goody bag, it should be folded into itself so it makes a small package, and carried loosely on a lanyard closest to the hand—so it can be dropped without having to remove your lights. Never have anything tied to you, especially your weight belt, and *never, never* use swing gate snap hooks, better known as suicide clips.

Some divers attach these brass clips to their weight belts or tank harnesses, so they can quickly snap in their lights, goody bags, or tools. This frees the hands, but the problem is that anything else can snap into it, too, such as the anchor line while making your descent, or cables inside a wreck. Once in, they can be literally impossible to get out, especially when positioned out of sight, forcing the diver to work by feel—usually with thick neoprene mitts. If you must use a snap hook of any kind, use either the brass snaps in which the slider gate is pulled up manually against spring tension, or stainless steel (aluminum corrodes) locking carabiners with a large gate. Even then, the snap should be on the piece of gear and the D-ring on your person, not the other way around.

Contrary to the advice of the certifying agencies with regards to weight belts, during a wreck penetration you do not want to lose your ballast. This usually results in your becoming overly buoyant, and inside a wreck control of buoyancy is important. Imagine snagging your quick-release buckle on a piece of wreckage, losing your weight belt, and soaring upwards, pinned to the overhead by too much positive buoyancy. Of course, neither do you want to be permanently attached to your lead.

The answer is the double buckle. Right next to your quick-release buckle, place another one, and run the belt through both of them. Then, if one buckle flips open, or if the hinge pin is bent out of place and lost, the other will act as a backup. For those of you who are afraid of the extra time, or thought process, involved in an emergency dump during the nonpenetrating part of the dive, I suggest you use only one buckle until you make your final equipment check immediately prior to entering the wreck. You can clamp down the extra buckle then, and release it later as you leave the wreck.

Check the rest of your gear, too: breathe off your regulators, switch on your lights, and check your air and bottom time. Look over your buddy; exchange okay signs; you are ready to enter.

Now let's discuss the techniques you and your buddy have decided upon. The simplest method, good for short penetrations, is the one in which one diver enters the wreck freely while the other stays at the opening shining his light in. The diver making the penetration should constantly look back, and never get out of sight. He should stop and turn back immediately should he not see his buddy's light.

For this you need a buddy you can trust: one who will stay there and concentrate on his duty, not one whose mind strays and who might wander off when he gets bored. This is a team effort. Later, you can switch roles, so each of you has the opportunity to do some exploring.

For deeper excursions you should consider the penetration line. Several types of reels are available, and there are different ways of using them. The outside diver can hold the reel and pay off the rope as needed, keeping a constant tension on it; then, when he feels it go slack during his buddy's return, he can pull it in. Or the inside diver can carry it for better control, either tying it off at the entrance point or allowing his buddy to hold a loop. Since it is possible to drop the loop accidentally, tying off is preferred.

The inside diver should not have the line tied around his waist, as in cave and ice diving. The main reason for this is awareness. If the line is looped over your wrist and gripped in your hand, you can have constant assurance that you are still tied off. However, as you move forward, the line naturally extends behind you and can easily become entangled in your legs, flippers, or high-pressure hose as you twist and turn. With the line in your hand, you can avoid the problem by keeping your hand well out from your body, and being careful as you change direction. Another advantage to this method is that, should the line snag on some projecting metal, you can free it more easily if you can see what you are doing and even transfer the line from hand to hand in the process.

The simplest reel is a spool or dowel with no moving parts, wrapped with a rope like sisal, a natural fiber that is cheap and expendable. It is unwound with one hand, by making circular motions with the wrist that allow the rope to peel off, or with two hands, by pulling off coils. This tends to be cumbersome, and is actually only recommended for spur-of-the-moment situations. Retrieval of the rope is complicated, often entangling both divers in the process.

A spear-gun reel is better. It can be attached to a short dowel, or right to your dive light, with plastic tie-wraps. The line can be spooled off by freewheeling, then reeled in as you work your way out of the wreck. The only problem here is that, because of the dimensions of the reel, only a thin line can be used. While this line is strong enough for pulling in large fish, inside a wreck, where it is constantly being rubbed against

A flight of stairs leads to a lower level on the *Texas Tower*. The profuse anemone growth means there is a good current flow, and that the steps do not lead into a dead-end room.

A hatch and staircase lead down to one of *Wilkes-Barre*'s lower decks.

sharp metal, it can be weakened by chafing or cut through completely.

Wire-tie reels, used by lathers and ironworkers, are about six inches in diameter and allow the use of a sturdy, eighth-inch manufactured line. The aluminum model tends to stick as the metal oxidizes, and needs constant attention. The high-impact plastic reel requires less maintenance: a shot of silicone for lubrication before each dive. Also, the bright yellow plastic stands out if the reel is dropped.

Yellow nylon line is strong, abrasion resistant, and highly visible. Polypropylene line has a drawback in that it floats: if you lose tension it can become tangled in overhead wreckage.

Now let's move into a big, roomy wreck where both divers want to go together. Start out by informing your friends on the boat of your intentions: where you plan to go in, how long you plan to stay, what method you plan to use. There's nothing worse than tying off your penetration line outside the wreck, only to have someone come by and slash it out of his way. Instead, tie it off inside the entrance point, but not so far from the opening that you cannot find your way out in zero visibility. Better yet, tie it in both places. If you are going in via a doorway, tie back the door so it does not close behind you.

The lead diver carries the reel; the follow-up diver runs his hand along the line and adjusts its position. When they return, the diver in front pulls the line free of any wreckage so the diver with the reel can pull it in. In some cases, where the divers expect to continue their explorations at another time, they may want to tie in the line permanently at the limit of their penetration. It should be pulled taut and tied high and clear of the deck so it does not get covered. When you use it again, you should check it for wear.

Despite the apparent security of the penetration line, it has its handicaps. It can break, come loose, get tangled, and literally tie up the user in knots. The worst hazard is that, as the diver executes turns inside a wreck or squeezes through collapsing bulkheads, the line may not follow his curving path. It can slide through notches in the metal, or slip between beams the diver has gone around. Then, as he feels his way back, he suddenly discovers that the line has taken a route he cannot fit through. If the visibility is poor, he may not be able to figure out where the line should go.

The moral here is orientation. Instead of a religious reliance on the penetration line, a diver should maintain a constant knowledge of his surroundings. I practice what I call "progressive penetration." The first time I enter a new area, I go no farther than a couple of body lengths. I study that small portion of the wreck, work my way out, then possibly do it again. The next time, I go over the same route, refreshing my memory of what it looks like, and extend my explorations another body length or two. In this manner I build up a series of recognition points, and memorize their relationships. I constantly know where I am, but I never become overconfident. At the first sign of trouble, I back out.

Fear is my greatest support, as long as it is under control. It tempers boldness, and keeps me from overextending myself. It is an extremely useful emotion, and those who do not have it, or who will not admit to it because of some insane sense of machismo, are in serious trouble. Know your limitations, and work within them.

The only reason a diver gets lost inside a wreck is because he cannot see to find his way out within the limits of his air supply. Rule number one is to start your retreat long before your air gets low. I would venture to guess that even in the most complicated, broken-down shipwreck a diver is never more than a few seconds, possibly half a minute, from the exit. He is not exploring half a mile of an underwater cavern system. Even a fairly intact freighter cannot be more than 500 feet long, a passenger liner a little longer. But the average penetration is like a fisherman's catch—it grows with the telling.

The greatest danger of penetration diving is silt. This is water-borne particulate matter created by the oxidation (rusting) of iron and steel, and microbial destruction of organic fittings: wooden decking and paneling, and cargo. It clings to every piece of wreckage and lies thickly on the bottom. While stirring up silt is unavoidable, the amount can be controlled.

Kicking hard with your flippers can agitate as much detritus as an airlift. You should avoid this mode of propulsion whenever possible. Instead, carefully neutralize your buoyancy, fold your legs up behind you, and push yourself along with an extended digit—unless it is a wide-open wreck, you will not have to counter any current. Literally poking your way through the wreck in this manner will greatly reduce the spiraling clouds.

Avoid dragging your hands, or body, along bulkheads, since that action will scrape off clinging rust. Do not play with the fish, since they too will kick up the silt. And do not get into frenzied bouts with lobsters. These can be found wandering freely in northern wrecks because the natural darkness satisfies their instinct for protection. The act of catching a lobster, opening your goody bag, and dragging the loose mesh on the bottom obliterates all visibility. Remember, the purpose of your dive is to explore the interior of the wreck—safely. Leave lobstering for another dive.

Caution should be used in picking up artifacts that are too tempting to leave behind. They will not get away from you, so you can take the time to ease them out of the silt and gently place them in your goody bag.

One thing divers rarely think about is descending silt: that is, flakes of rust dislodged from overhead by your exhaust bubbles. As you turn around you will see these large particles dropping and obscuring the visibility. It is wise to keep in mind that no matter how careful you are, silt will always be present to some degree—and it will always be worse on the way out.

Despite all these precautions, suppose you do get lost. What is your best course of action? "Don't panic" goes without saying. Personally, I always panic right away—that is the initial reaction of fear. But I fight it down, stay still, concentrate on controlling my breathing rate, and think.

Getting lost usually means losing orientation. If you have used progressive penetration effectively and have examined your route, you have a good chance of getting reoriented simply by placing your light close to the deck and seeing which way the ribs and cross-members run.

In the ammunition room on the *San Diego,* two ocean pouts stand guard over clips of bullets for World War I-vintage Springfield rifles.

Kenny Gaskin poses with a compass binnacle recovered from the *Malchace.*

Resist the temptation to spin around in circles, because then your orientation is utterly gone. If you need to turn around to head out, place your hand on the deck and make a 180-degree turn while watching the bottom. Experiments have proven that almost no one can execute a perfect about-face blindfolded.

Without overusing your flippers, come to a standing position in order to climb above the largest percentage of silt. Put your light against your chest (*never* turn it off) and look around for that green glimmer of light seeping in through cracks, portholes, and other openings.

Look for marine life. Filter feeders like anemones need to be near the current that carries their food supply, and are not plentiful deep inside wrecks where the water is relatively stagnant. Big ones, or clusters of them, will lead the way to an opening—although not necessarily one large enough for you to fit through. They can, however, point the way along a corridor as they become more and more abundant.

If you do become hopelessly lost, moving in circles will not help. Pick out a direction and stick to it. Psychologically, ducking into the silt is difficult to do—yet the silt is most likely the trail you made on the way in, like the bread crumbs of Hansel and Gretel.

Once I even used a compass inside a wreck. Granted, pinpoint accuracy was not possible due to the surrounding iron, but magnetic lines of force do exist. I knew the axis of the wreck was north-south, so I followed the needle west until I ran into a bulkhead, then swam along it until I came to the opening.

The more knowledge you have of a wreck before you go inside it, the better equipped you are to handle a given situation. But more important than this is to know yourself. Wreck penetration is not for beginners. It is as methodical as a science. You must first be secure in your diving skills before undertaking this great step. Then you must be willing to take the time necessary to gain the experience required for this specialty.

And finally, you must not be afraid to be afraid.

CHAPTER 4

Deep Diving Procedures

Deep diving has often been described as long hours of boredom punctuated by moments of sheer terror. Obviously, for one who is unaccustomed to depth, it is easy to be intimidated by the inherent dangers. Those who regularly dive deep agree that the most hazardous situations are encountered on the highway to and from the dock.

Like any other high-risk sport (skydiving, skiing, and mountain climbing), deep diving involves special safety precautions, learned through training and experience, which prepare the individual for possible difficulties. So, rather than decry the activity, let's take a positive outlook and learn how the sport can be made safer.

The most important aspect of any diving is air. In deep diving, one must not only be assured of an adequate supply, but also have an emergency backup: the surface is too far away to make a free ascent. Obviously, a single tank with one regulator will not suffice.

The first step is to move into a double-tank setup. The cubic inch displacement of contemporary scuba tanks ranges from 50 to 94. Twin 50s are lightweight and easy to handle, but hold only slightly more air than a single 94. For all but those few who have fantastic breath control, they are inadequate. Twin 94s are great for air hogs if they can manage the excessive weight and have a large capacity for buoyancy. But for most people the optimum size is either twin 72s or twin 80s, the 80s being preferable.

The way doubles are rigged is a point of controversy, usually left up to personal preference. The simplest setup is to use two single tanks in a double harness, with one regulator on each tank. Besides utilizing equipment that most active divers already have, this system has the advantage of lightness while ensuring an alternate air supply should one tank blow a disk or a regulator fail. In addition, if only one tank is used, only one tank needs to be refilled. In that case, a third single can replace the one used tank for a repetitive dive. However, the disadvantage is that the diver must anticipate switching regulators during the dive. Ideally, he should keep a constant check on his gauges and make the switchover before completely draining one tank, not just so water does not get sucked into the empty bottle, but so it can still be used as a backup. Invariably, the diver fails to make the changeover until breathing becomes difficult, putting him in the precarious position of *having* to switch regulators, regardless of circumstances, on every dive, usually when he least expects it. This means there is one more thought the diver needs to hold in his head. Also, the regulator on one tank needs to have a special long hose in order to allow free movement of the head.

The next step up from this is a crossover bar, which yokes together the valves of two single tanks, leaving one common point for the regulator. The crossover bar is an inexpensive piece of equipment which can easily be removed if the diver decides he also has use for single tanks. It eliminates one regulator. And it is just as light as the two-singles system.

A cold water wreck diver in full regalia.

However, the diver loses the advantage of a backup regulator. Also, a hard drop on the deck may dislodge one or both of the seats, or twist the bar. And a blown disk during a dive causes all the air to be lost, instead of just half. The major problem seems to be that, in the rush and tumble of getting dressed, sometimes a diver may neglect to open one tank. The pressure gauge will read full, but only half the supply is available, resulting in dives that are terminated in a panic unless the diver has a long reach, or his buddy can act fast enough to open the valve.

More stable than this arrangement is the permanent double yoke, which adds stability to the double-tank setup while adding only a little extra weight. Also available are double yokes with two regulator ports, so a backup regulator can be used. These are expensive, but afford the diver a completely separate means of getting air out of the tanks in case of emergency. If his main regulator fails he can use the spare; if his buddy's fails he can pass him a second mouthpiece.

Some divers prefer an octopus regulator: that is, one first stage mounted on the tanks, with two second stages. This is only half a backup, since an auxiliary second stage is of no use if the first stage fails. It is best kept for instructors who have a constant need to offer their own regulator to a panicked student.

The ultimate backup is the pony bottle. This is a scaled-down cylinder offering an additional twenty to forty cubic feet of air, which it delivers through its own regulator. It is mounted with stainless steel hose clamps in the curved space behind the doubles. Besides being an alternate air supply, it is also an additional air supply. It is important in this case, as with any extra regulator, to have the second stage firmly secured. It cannot be left dangling, for when the user wants it, it will likely be out of reach, behind his back. A lanyard with a clip can attach it to the tank harness or BC. A neck strap will ensure that it is directly under the mouth. It is a good idea during a dive to grasp the spare second stage every once in a while so you are sure it is readily accessible. And test it periodically so you know it is working.

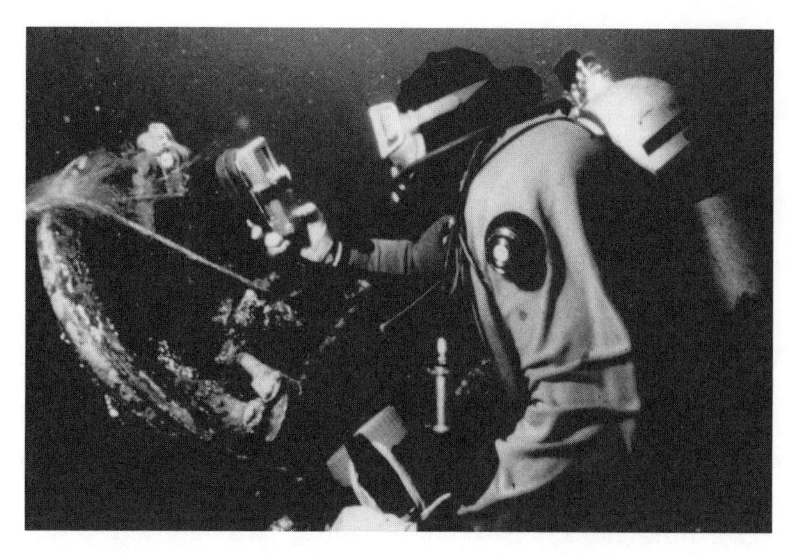

Larry Keen checks his gauges before crawling down the after torpedo loading hatch of the U.S. submarine *Tarpon.*

Even the novice diver understands that, by simple extrapolation of Boyle's law, a greater volume of air is consumed the deeper the dive. What is usually overlooked is that, at depth, the slightest exertion causes a geometric increase in air consumption. Many regulators on the market cannot deliver the quantity of air necessary for a fast-swimming or hard-working diver at 200 feet. The diver quickly "overbreathes" the regulator; that is, he demands more air than it can deliver. This puts the diver in the painful position of gasping, or sucking in as much air as he can get.

This is a dangerous situation. First of all, if there is any small emergency, the diver cannot meet it calmly: he is concentrating on the more basic need of getting air. If his supply is low, he is using it up at a prodigious rate. And if he is hyperventilating, he is in extreme danger of passing out. It is not the lack of oxygen but the buildup of carbon dioxide, a waste product of muscle exertion, that triggers the breathing mechanism. The fast, shallow breathing of hyperventilation blows off the carbon dioxide without bringing in sufficient oxygen, causing unconsciousness.

Some diving casualties are found with full tanks and functioning regulators, with no apparent cause of death. I once saw a diver lose his fin trying to kick off the bottom at 185 feet. Instead of inflating his dry suit, he kicked harder. In moments he blacked out. Fortunately, his two buddies saw him sinking back down, went back for him, grabbed him under the arms, and swam him up. He regained consciousness at 60 feet. If the circumstances had been different he would have become another statistic: a diving fatality for no known reason.

Besides the obvious shortness of breath, a warning of impending loss of consciousness is dizziness, nausea, or tunnel vision. Under these circumstances the diver should immediately stop what he is doing, grab onto something, and breathe as deeply as possible until all symptoms abate. If you really think you are going to pass out, make yourself positively buoyant and start heading up. Then, even if you do black out, at least your unconscious body will appear at the surface where help can be gotten. Remember that scuba uses demand regulators, and that only a conscious diver can cause the muscle contraction that forces the lungs to expand and make that demand. An unconscious diver cannot make the regulator provide air to his body. As in most cases, prevention is better than a chance for cure. Never get yourself into a position of overbreathing. Instead of swimming hard, just do not go as far. If an artifact will not come off easily, leave it alone instead of beating on it further. No artifact is worth the price of your life.

Buoyancy control is important in deep diving. In addition to providing more than ample buoyancy, a dry suit also keeps the diver warmer, thus slowing down his breathing rate and conserving air. In emergencies, a simple flick of a finger can drop thirty or forty pounds of lead in the blink of an eye, making the diver that much more positive.

If you are using a wet suit you must have the largest volume buoyancy compensator available. If it does not come with a crotch strap, add one to keep it from strangling you or floating in front of your face when inflated. And be careful not to overweight yourself. Extreme depth crushes neoprene, so that your suit loses buoyancy at the same rate it loses insulating quality.

As the Romans often said, *tempus fugit*. Time goes by quickly on a 200-foot dive, and one must be constantly aware of air, bottom time, depth, and decompression penalties. Gauges should be checked frequently. I can remember (vaguely) a dive to 200+ feet during which I once looked at my gauge panel, put it down, then realized that I had no idea what any of the instruments showed. Fighting narcosis is like trying to stay awake during the late late show. Somehow, you are never really sure what's going on until the test pattern wakes you up. When diving deep, you cannot just look at your gauges, you must study them in order to interpret the readings.

Gauges are best clustered on a panel, where you can see them all at a glance. Because your brain may be numbed by narcosis, they must be easily understood. I once had a depth gauge which was reading only 30 feet when it should have said 230. I had to assume it was broken and rely on the planned dive profile for depth and decompression calculations. It was not until after the dive, in a clearer state of mind, that I deciphered the meaning of the needle's position. It had swept around one time and was reading on an inner scale. This kind of gauge cannot be tolerated in deep diving: the information must leap out at you.

Bezels are for the birds, and second sweeps keep getting confused with minute and hour hands. Overcoming this, the Bottom Timer is easy to read and will work at any depth attainable by scuba; and by the time it reaches the zero point and starts counting another sixty minutes, you had better be on the decompression line.

But the new generation of digital readouts really fills the bill. There is no confusing three o'clock with quarter after twelve, and you do not have to remember what time you hit the water. When the pressure switch actuates at four or five feet and your timer says seventeen, you can be sure that you have been down for seventeen minutes. At extreme depth, every minute, every second, is critical when your air gauge is dropping like a rock and your decompression meter is taking off like a rocket. *You must know your time.*

Bottom time is so important that I always carry two timepieces. If they do not synchronize at any time during the dive, I take the more conservative approach.

Which brings us to the reason for the precise knowledge of bottom time: decompression. While the science of decompression is a topic all to itself, it must be understood that in deep diving an extra minute on the bottom can be the difference between being straight and getting bent. If you plan a twenty-minute dive to 200 feet, you must be ready to ascend at twenty minutes. Of course, you should also keep in your head a contingency plan in case you overstay your time. This is the reason for the pony bottle and its extra air supply, and a backup timepiece for use with the *Navy Tables.* Everything you do should be planned. If you are simply exploring a deep wreck and have no plan other than exploration, make sure you have with you the necessary equipment to make a remote decompression.

Kenny Gaskin peers into the after hatch of the *Tarpon*.

In other words, be self-reliant. Stay with your buddy if possible, but never rely on him. A buddy is not a crutch to make up for your weaknesses; he is an added safety factor, a backup.

If you get entangled in monofilament and drop your knife in your eagerness to cut yourself out—have a spare knife. If you are going to swim out of sight of the anchor line on an unfamiliar wreck in poor visibility—have a compass. And if you get into trouble of any kind, be prepared to head for the surface—immediately—no matter what the consequences. You do not have the air supply to finesse your way out of a delicate situation.

Getting bent is better than getting killed.

That is why your gear must *never* be tied to you. Sure, you don't want to lose that expensive camera—but your widow could always buy a new one. Sure, you worked hard for that lobster, or porthole, but you do not want the extra weight to nail you to the bottom. If you must clip something to your weight belt, make sure your clips are well maintained and work easily. Better yet, carry your cameras or goody bags on a lanyard hung loosely over your wrist; in an emergency they can be dropped easily.

And never—never—go beyond your limits. Progress into deep diving as you would advance into expert skiing. You do not jump off a vertical bluff the first time you have fiberglass under your feet. Choose your destinations so that each one is slightly deeper than the last. Make sure you are comfortable at 90 feet before diving to 100 feet. The only way you can test your tolerance to narcosis is to sneak up on it.

Sure, all this takes time. But if you do it right, you have the rest of a long life to do it in.

CHAPTER 5

Decompression Methods

According to most instructional agencies, decompression diving does not fall within the realm of sport diving. It is something to be avoided or ignored. Yet there are many divers who commonly make decompression dives and think no more of it than they think about eating a sandwich during surface interval. If they are not sport divers, what do we call them? Certainly, they are not professional divers or commercial divers. They are diving for sport, but there is no easily definable term for them. They are not decompressing in order to maintain a macho image, and most of them—perhaps all of them—hate decompressing: it is simply the price they must pay for extended bottom time. If anything, they can be called decom divers. Therefore, I will let those who enjoy labeling conjure up a name for them. To me they are simply scuba divers.

The question this raises is: how does one become a decom diver?

Consider an incident which occurred recently off the New Jersey coast. Seventy miles at sea, a group was making a dive on a sunken freighter in 190 feet of water. One diver, obviously a novice at decompression diving, became disoriented on the bottom and could not locate the anchor line, on which he was planning to ascend and decompress. He had to come up free and do a floating decompression, using only the buoyancy in his dry suit to maintain the proper depth. This is almost impossible to do with any accuracy and should only be considered in an emergency. In this case, of course, it was an emergency.

In the thirty minutes he took to complete the required stops, the current carried him out of sight of the boat. When neither he nor his bubbles appeared on the surface, those on board thought he must be dead. The boat captain, himself a diver, went down to look for the body. An hour later he returned empty-handed.

He was about to call the Coast Guard when a fishing boat radioed that it had picked up a diver floating in the ocean—alive, and in very good condition. The boats rendezvoused, and the diver was transferred back to his own charter boat. The probability against being spotted by a passing boat seventy miles at sea is astronomical! He was lucky.

At least this diver knew he had to decompress. Perhaps more common is the case in which a diver plans a no-decompression dive and forgets to look at his watch until he has exceeded the limits. Or he is just about to leave the bottom when he becomes entangled in monofilament and takes several panic-stricken minutes to free himself. Now decompression is required, and he has no idea how to go about it, or how long he must stay, and at what stages, before he can come up safely.

Obviously, what is needed is more education in the art of decompression. Divers need to be familiar with the physiology of the bends, recognize the symptoms, understand the treatment—and know how to decompress.

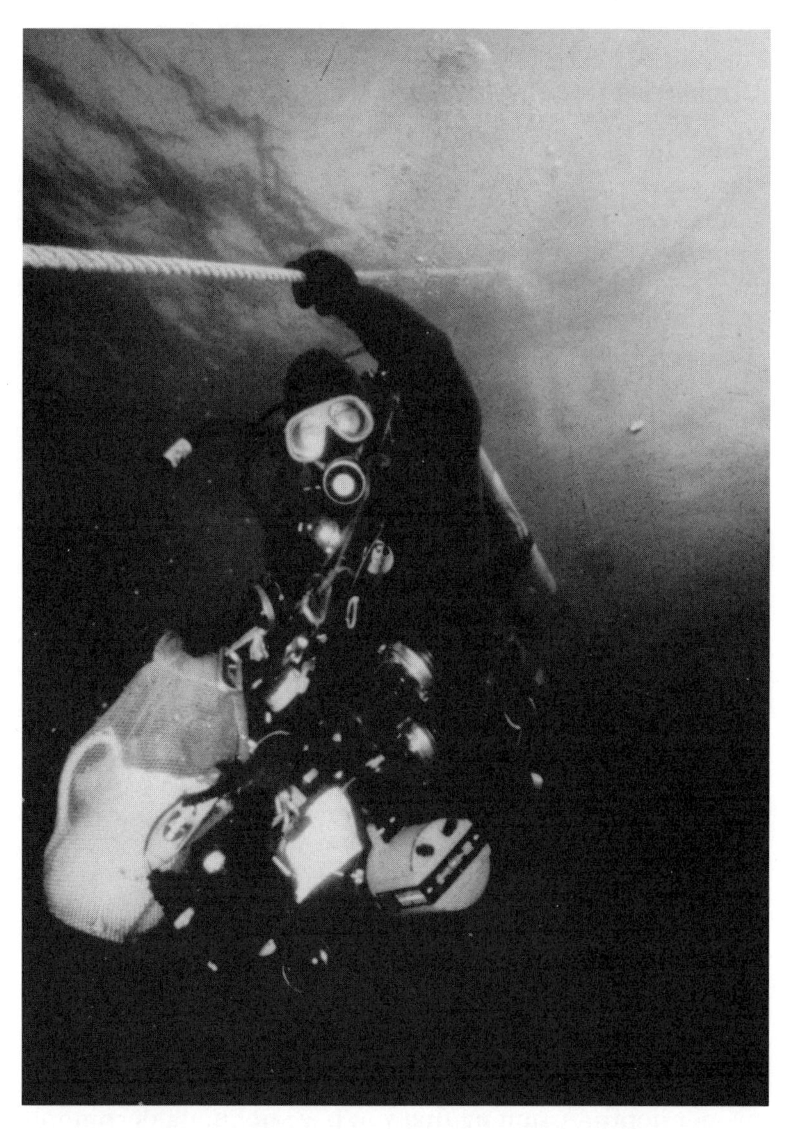

With camera gear and a china pitcher, recovered from the *San Diego,* Jon Hulburt waits out a lengthy decompression.

Unplanned Decompression

Many divers have never decompressed, and never plan to decompress. In dive class, the *Navy Tables* were a jumble of figures and a mass of calculations that were all but incomprehensible. They may have learned enough to pass the course. But several years down the line, all knowledge of reading past the no-decompression limit for a specified depth is but a hazy memory, and repetitive dive groups and residual time an arcane mystery.

This leaves the diver in a serious position when the unexpected occurs, and he finds himself too deep for too long. The initial reaction is fear and panic, and the novice will scream for the surface with an instinctive but false perception of the safety it seems to offer. It is only the level-headed diver who will acknowledge that, although his position seems untenable, he does have viable options and can avoid the dreaded bends.

It may be academic to state that, as soon as this situation is recognized, one should immediately head for the surface. But it is extremely important to realize that too quick an ascent can bend a diver just as drastically as missing a stop. Start up, but follow those tiny bubbles.

If the anchor line is nearby, and you can see it, go for it. But don't waste time on a prolonged search, using up valuable air and increasing your decompression penalty. If possible, signal your difficulty to another diver (other than your buddy). There is no hand sign for this, but the universal finger across the throat declares trouble and should elicit aid.

You are faced with two immediate problems: how long to decompress and at what levels, and how to maintain the proper depth. Assuming that you have neither a decompression computer nor a set of *Navy Tables,* you are left to your own devices. A glance at the *Tables* will show that there is an inordinate time differential between the no-decompression limits and the time requiring a twenty-foot stop. For example, a planned no-decompression dive to 90 feet extended from

During a lengthy decompression on the anchor line, Dave Stroh studies his gauges to check proper depth and time. The spool of line clipped to his belt is an emergency decompression reel, to be used in case the anchor line breaks free or cannot be found.

thirty minutes to sixty would still not require a stop at 20 feet. Five minutes at 150 feet would need to stretch to fifteen minutes. It is unlikely that anyone planning a no-decompression dive would stay down twice to three times as long as planned. Therefore, it is safe to assume that in most cases a stop at 10 feet will suffice. If you are uncomfortable with this, make an additional stop at 20 feet for two or three minutes. It can't hurt.

To extrapolate the time at 10 feet I use the rule of halves. If you stayed down half again as long as you should, at 90 feet the decompression penalty would be fifteen minutes. The rule of halves at 80 feet would require seventeen minutes, at 70 feet, eighteen minutes. Everything else is downhill. At 150 feet, a stay twice as long as the no-decompression dive would require only one minute of decompression.

Since the no-decompression limit is likely to be the only number in your head, I will sum up the rule thus: *If you halve again your bottom time, decompress for half your no-decompression limit.* Any diver so unobservant that he exceeds his planned bottom time by more than fifty percent shouldn't be diving—or chewing gum.

Now that you've picked your level, you have to stay there. Free-floating buoyancy control is difficult at best, but it can be done. Keep your eye glued to your depth gauge, stay slightly negative, and use a slight waving motion of your fins to maintain depth. If you get even the least bit positively buoyant you're likely to rocket to the surface before you can dump the excess air.

Some divers weight themselves in such a way that they literally cannot get under without a line to pull themselves down. This is as dangerous as being too heavily weighted. They lose the option of hovering. The name of the game is neutrality.

Face down-current. That's where the dive boat will be, at the end of its scope, so there is a chance you'll see the anchor line as you drift by. And while you're doing all this, take out your knife and continually rap on your tank. The sound will travel underwater, and someone might get the idea that a diver is in trouble somewhere. At least he can pass on the word, and those on the surface can be on the lookout.

If you've dropped your watch, or it has stopped working, the old one-thousand-and-one count can approximate your decompression time. A twice-per-second knife beat can help in accomplishing both aims. Barring this, I subscribe to the rule "when in doubt, hang it out." If you've lost all track of time, just breathe slowly, deeply, and until your tank is dry. A single minute of decompression may be the difference between paraplegia and a happy ending.

Suppose the worst happens, and you run out of air. Need I say, come up? Don't hover until your last breath is gone. When you feel the first difficulty in breathing, get your head out of the water, swallow your pride, and yell for help. I've seen situations where a diver, watching his tank pressure gauge closely, broke decompression with a hundred psi left, or halfway through his pony bottle, called for more air, then dropped back down. Surface personnel swam out a spare tank on a line and lowered it so the diver could complete his decompression. The short bob to the surface did not materially affect his penalty, although at the very least an extended stay was called for.

Believe it or not, before sport divers started carrying their own decompression reels, they commonly used to break the surface, swim to the boat, and go back down the anchor line—with no ill effects. While this is not recommended, the onset of the bends normally takes several minutes. For decades it was standard operating procedure for the U.S. Navy to yank their hard-hat divers out of the water, speedily undress them, and chuck them into an onboard recompression chamber. I don't care who does it, I still don't like it. But it usually works.

Still, if you surface short, immediate recompression is the safest course to take. Don a full tank and get back down. In all predicaments where decompression is interrupted, when a diver redescends, someone should get in the water with him and keep up a constant vigil. In the case of a severe cerebral hit, he could lose consciousness and sink. Psychologically, too, having a companion to watch over him is calming, allowing him to make better decisions about how to proceed.

Assuming you've done everything you can in the water, including buddy breathing, you can still counteract decompression symptoms on the surface. Most dive operations carry O_2 bottles, so as a precautionary measure you should breathe pure oxygen for at least a half-hour. Remove constricting clothing and increase your awareness of your body; feel for tingling sensations, pinch for numbness. Pain will soon become apparent.

If you feel out of the ordinary, keep your buddies apprised of your condition. As in hypothermia, once you start suffering the symptoms of caisson disease, you are no longer capable of making decisions. Fear of ridicule, of inconvenience, of helicopter rides, of expense, can influence your judgment. Let others decide for you the proper course of action: to contact the Coast Guard, to arrange transportation, to prepare the chamber, to have doctors on hand. All organizations that can offer aid should be notified, just in case.

At this point you've done all you can. Even if it develops that you do not need those services, you should make a full report of your activities to authorities who can use the information constructively. At the very least, please learn something about yourself, and don't let it happen again. You can be helped against ignorance or accident, but not stupidity.

Planned Decompression

The *U.S. Navy Standard Air Decompression Table* is the bible of decompression diving (also see "Caution" in chapter 2). Study it until you know it forward and backward. Technically, footnotes make it self-explanatory, but the concepts are easier to grasp when presented by an instructor. In addition, there are several good books on the subject.

To follow the *Tables* requires rising at a controlled rate of sixty feet per minute, and stopping at prescribed depths to allow the nitrogen in the tissues to come out of solution and be expelled from the blood before expansion and bubble formation constrict circulation. The common way to do this is to come up the anchor line. The diver can monitor his watch and depth gauge, while controlling his ascent by holding onto the line all the way to the dive boat.

There can be a few problems. The diver must first locate the anchor line on the bottom. Even with a good sense of direction, and a compass, finding the line in poor visibility is not always easy—or possible. Even in clear water, or when the wreck is small, or laid out in such a way that orientation is made simple, there is always the chance that the anchor line may not be there when the diver returns.

Anchors and grapnels have a habit of pulling loose, especially when high seas jerk the line taut, ripping the anchor or grapnel out of the wreck or straightening the tines. On smooth days without waves or current, the grapnel can come loose through the slack in the line, and the boat can drift away. The solution here is to "tie the hook": the first diver down takes the grapnel and snags it in a good spot; or he wraps the chain around a piece of wreckage; or he takes another line and uses it to tie the grapnel to the wreck. This takes coordination with the captain, since he must keep the boat moving slightly forward—enough to counteract the current, so that the line is kept slack while the diver performs his task. When the job is done, communication is necessary so those on the boat know when it is all right to go in.

This is done in a variety of ways. The tie-in diver can return to the surface and make a verbal report; or he can jerk on the line to let the mate holding the other end know he is finished; or he can simply not come up, and after a reasonable amount of time elapses and the line is tested for tension, it can be assumed that the job is done. He can also use a GSCD—the grapnel set communication device. This ingenious invention is a styrofoam coffee cup which the diver carries down with him, in a goody bag or BC pocket. Upon its release it floats to the surface, where it is spotted by divers waiting to hit the water. The acronym is pronounced "gascod," but the device is usually referred to simply as "the cup." For deeper dives, a marker buoy inflated by a CO_2 cartridge can be used.

There are problems and inconvenience with anchor-line decompression. The worst is crowding. Once I was diving on a wreck in 120 feet of water in 100-foot visibility. This caused two unpleasant side effects: everyone overstayed his time due to the unusual clarity, and everyone found his way back to the anchor line. For twenty-five minutes, eighteen divers fought to get their hands at the 10-foot mark. Fortunately, it was a calm day. But there are always small mishaps occurring: masks getting knocked askew by flippers and dangling crowbars, gear bags getting intertwined, and pressure gauges catching on the brass clips of another's weight belt. To make matters worse, someone sent up a liftbag clipped to the anchor line, and divers at the 20-foot stop were peeled off by the rising carabiner. To my great shame, I was the culprit. I endured many angry stares until some of the divers found they could hang onto the line from which the hundred-pound artifact was dangling. I was later absolved of crime.

Buoyancy can also be a problem. Ideally, the anchor line should be as straight as a firehouse pole. But unless the seas are flat this puts too much strain on the boat, the line, and its point of connection. Yet, the more scope in the line, the more difficult it is to hang at the right level. As the boat rides up and down so does the diver, alternately placing him either too deep or too shallow.

A diver whose dry suit or BC is slightly overinflated will cause the line to lift, while one who is negatively weighted will cause it to fall. Thus, there is a gyrating effect as the divers turn the line into a yo-yo. The more slack in the line, the worse this condition becomes, until a point is reached where the line looks like a zigzag lightning bolt, where divers closer to the boat are deeper in the water than those supposedly lower down. Then, when the whole contrivance rises to the surface, there is mild panic while everyone scrambles to a lower position which, because of the sudden weight, plummets twenty or thirty feet. Then everyone dances back up.

In rough seas this exaggerated up-and-down motion can easily make a queasy diver seasick. Having had to vomit through my regulator on more than one occasion, I can state unequivo-

Decompressing divers demonstrating the use of the jonline.

cally that this is an uncomfortable way to decompress. It's also bad on the shoulder joints, as your arms are nearly ripped out of their sockets. The trick to maintaining the proper buoyancy for your stage is to hold the anchor line lightly with encircled thumb and forefinger, so it slips up and down through your hand. I once wore a hole through my mitts during a single hang, but at least I was not bounced around like a basketball.

If you anticipate such unmanageable conditions, a better solution is to carry a jonline, appropriately named after its inventor, Jon Hulburt. This is a length of rope three to six feet long, with hand loops at both ends. Snug it around the anchor line like a choker by pushing one loop through the other. This will hold it at the proper depth while you hang onto the other loop. Now, as the anchor line jerks up and down, so does your knot—but not you, three to six feet away. With your buoyancy set, you remain stationary. Using a six-foot rope with an additional middle loop gives you a choice of places to hang on, depending on wave height. You will find that this system works extremely well in strong current, too, because you have a firmer grip on a loop than on a slippery, angled nylon line.

Jon goes one step further with his method. In order to have his hands free during the dive, he ties a short line to his tank manifold and runs it over his shoulder so it dangles at chest height. A locking carabiner on the bitter end acts as an equipment ring. He snaps on each light and goody bag with its own carabiner. During decompression he clips in the jonline, then lies back and relaxes his arms until his bad time is up.

The question we must now address is: how does a diver insure a safe method of decompression when he suddenly finds himself preparing to make a deep or extended dive without an anchor line, or when he cannot rely on relocating it?

Let's take the hypothetical example of a diver visiting a Caribbean resort and having the chance of making an over-the-wall dive, something that is done every day. The average diver, offered the opportunity, will not turn it down—the retelling of the adventure often acts as its own incentive, and thoughts of safety are scoffed at and put aside because of the great need to appear bold. Vanity brings about more injuries and deaths than any other cause.

The first decision the wise diver must make is not "Can I do it?" but "Is it safe—for me?" In the first place you are judging not your ability to make the dive, but the conditions under which the dive is to be made: wind, current, visibility, and back-up safety features provided by the diving concession. Is the boat adequate? Is the captain knowledgeable? Is the crew proficient? Is the dive master competent?

Assuming the answers to all of the above are yes, there is one thing you must realize: you may not yet have gained the necessary expertise for such a dive. Many people make the mistake of relying on the experience of the dive master to get them out of trouble, since he leads the group or the individual on the dive, taking him down by the hand and bringing him back up again. Here I must stress reliance on one's own ability, since it is easy to get separated from a dive master or buddy. It's nice to have someone along for emergencies, but not as a crutch. If you need such a crutch, you should not be making that dive. You should never dive beyond your own ability. Instead, gradually better yourself through continued training, either on your own or through an agency.

Self-confidence is your best buddy. A dive that is safe for someone else is not necessarily safe for you. Do not be trapped into forcing yourself beyond your limits by peer pressure, or exuberance. Do not be intimidated into making a dive which is beyond your present capabilities. This is where intelligence should take control, where maturity should show itself, where rational thought should overcome your emotional whims. You are playing a game in which your life is at stake, and this is not to be taken lightly.

But once you have reached the stage in your diving career where you feel comfortable making a decompression dive, what can you do to further reduce the odds of unforeseen accident on your resort dive?

First, find out where the nearest working recompression chamber is located, and what provisions are available for getting you there quickly. On one occasion, a diver suffering from a severe case of the bends was flown from one island to another in an unpressurized plane—he died on the way.

Next, talk with the dive master and boat captain about what kind of emergency equipment is provided. Some tropical resorts use a length of metal or PVC pipe, suspended below the boat on two ropes, like a very wide swing, affording plenty of elbow room for decompressing divers. One boat captain in the Florida Keys furnishes a complete oxygen rig with half a dozen second stages feeding off an on-board bottle. Breathing oxygen while using the *Navy Air Decompression Tables* offers an increased safety factor, providing one has had an oxygen toxicity test. In lieu of that, a hang bottle filled with air and enough regulators for all the divers in the water should be tied off to the boat at all times. And an oxygen first aid kit can greatly alleviate any post-dive symptoms until a chamber can be reached.

Finally, what measures can the diver take for himself? At the very least he can buy a spool of sturdy string from a local store. This can be rewound on any kind of board, preferably a dowel, and stuffed into a goody bag. With a thick leader of rope, say two or three feet long, the end can be tied to a convenient coral head and unreeled to the surface, allowing the diver to maintain his depth during decompression. For diving in sandy areas, a five-pound lead weight can be tied to the end. When deposited on the bottom, it will give some form of stability while the diver makes his ascent.

Another method is to take along a pop buoy with forty or fifty feet of line attached to it. The diver can then make a free ascent, using his BC to control his rate, and as he approaches his first decompression stop he inflates the buoy while holding onto the end of the line. His buoy then acts as a surface marker for the dive boat, and he can float along underneath it while keeping his proper depth. All these methods, however, are emergency procedures only, and are not to be used as a matter of course.

Enter the personal decompression line. Under its many pseudonyms the ascent line—up line, down line, swing line,

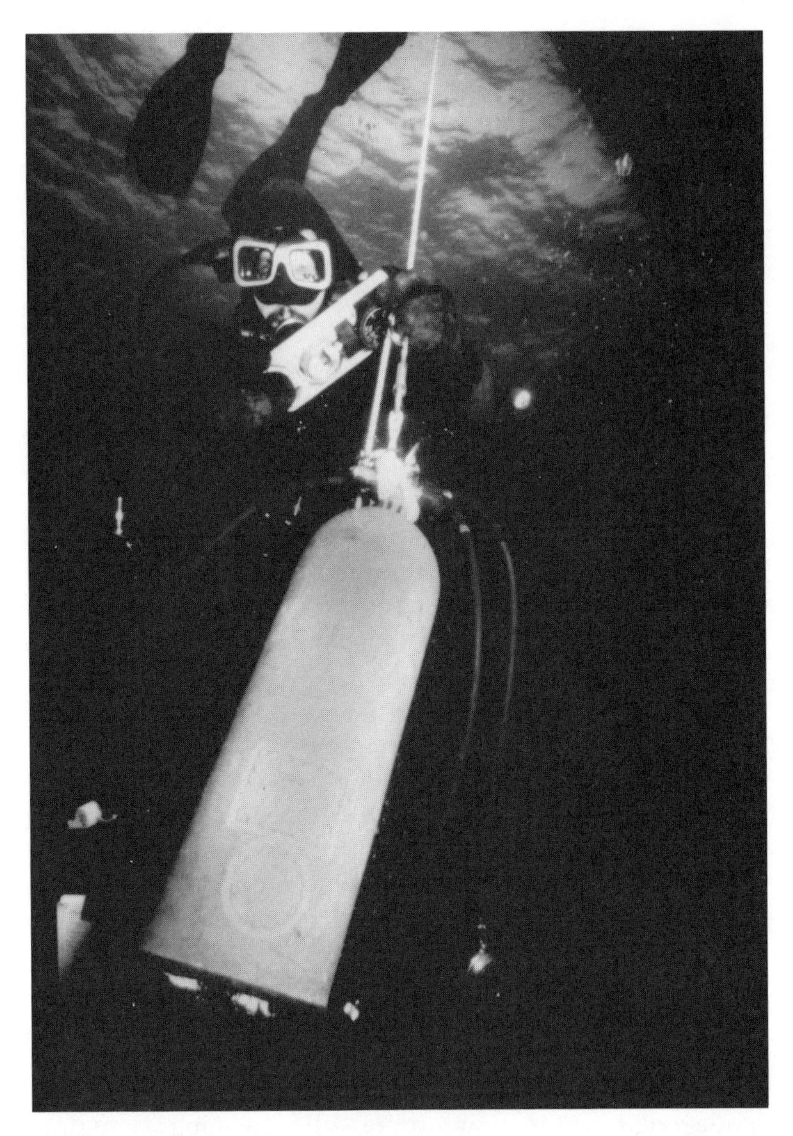

A spare tank complete with octopus rig hangs below a dive boat, with extra air for decompressing divers.

hang line, decom line—is simply a line the diver carries with him, ties off to the wreck, then unreels as he rises, using it to maintain depth while decompressing, without floating off into the sunset.

In its simplest form it is a piece of strong string carried on a spool. A rod setter's wire-tie reel, made out of plastic, can be wrapped with eighth-inch nylon line. When the diver leaves the bottom he pulls out some line, ties it to a convenient piece of wreckage, and holding the reel in his hand, pays out line as he makes his ascent. When he reaches his decompression ceiling, he hangs onto the line to maintain the proper depth. For speed, a large, expendable metal clip, looking like a clothespin, can be secured to the end. Then, instead of having to tie a knot while on the bottom, the diver merely loops the line around the wreckage and clips it on itself. For abrasion resistance, a metal leader can be used. This type of reel is small in size and weight and can easily be carried in a goody bag. However, make sure that a bagful of lobsters does not make it inaccessible.

An added cautionary note: When tying off on a piece of wreckage make sure that the wreckage is firmly attached and will not break from the strain, and that it is not open-ended, thus allowing the loop to slip off. Keep clear of sharp metal projections that can cut the rope. And be sure of your knots.

A larger version of the personal decompression line is the broom-handle model, which has a reel consisting of a 2-foot-long wooden dowel 1½ inches in diameter. About 4½ inches from each end, drill a ¼-inch hole and insert a 6-inch metal dowel. (Wooden dowels will snap off when the submerged rope gets wet and expands.) Now there is approximately 14 inches of space on which to roll cheap, expendable, and biodegradable ¼-inch sisal rope, with a breaking strength of 600 pounds. Each wrap is longer than the last, so that the total capacity of this reel is over 300 feet. Also, it is suggested that 5½-inch circular plates, either plastic or wood, be used on the inside of the metal dowels to prevent the wet rope from bulging; otherwise it will snarl as it is unreeled.

After a dive on the *Wilkes-Barre,* divers complete their thirty-foot staged decompression before moving up to the oxygen-fed regulators waiting at twenty feet.

Whereas the wire-tie reel can be managed with one hand, the broom-handle model requires two. It is also long and heavy, and carrying it can be tiring, especially when the goody bag in which it is held snags on every sharp piece of wreckage. For this reason, it is usually worn on the back or the side of the tank(s), and held in place by two thick rubber bands or sliced automobile inner tubes that stretch around the tank(s). The diver then reaches around behind him, thumbs off the lower tube, and drops the reel out of the upper tube. This requires some practice, so one should test out his rig and gain confidence before real use. But at least it is out of the diver's way until it is needed.

In order to ensure that a tank-carried decompression reel is not lost, either from a broken band or by getting snagged in wreckage or netting, many divers attach a lanyard to the lower end. A clip from this can be snapped onto your tank harness or BC strap—but *never* to your weight belt. This way, if the reel is dislodged, it will dangle by your legs, and kicking it will alert you immediately that it is loose.

Ascending can be a little tricky. With both hands occupied holding the reel, the diver must stop to release expanding air from his dry suit or BC, while not letting out too much, which would cause him to start dropping. He must not get the rope caught in his flippers, belt clips, or goody bag. He cannot see his gauges, so he must watch his bubbles to measure his ascent rate.

A more complicated reel to make is one in which the two-foot dowel is replaced with a length of aluminum pipe. A stiffened metal wire perhaps an eighth-inch thick is inserted through the hollow ends and bent into a rectangle that reaches far enough beyond the wrapped rope that the diver can hold onto it with gloved hands. Now he can manage unrolling his line with one hand.

The worst problem with this method is the effect of the current. Once you've hung on an anchor line like a flag in a gale, you know you could never have the strength to do it on a freely swinging line. Because of the angle produced by the current, you may need 200 feet of line to reach the 10-foot stop in 130 feet of water. Then, because of the pendulum effect, the current will push you down deeper than your stop. You will add air for buoyancy, and rise. It will expand and push you up too high. You will release air, and the current will push you too far down. Besides being unsafe, this yo-yo effect is exhausting.

The alternative is to tie a liftbag onto your decompression line, inflate it, let it rise to the surface, then cut the rope and tie it off. Now you have your own, personal anchor line which can be easily watched on the surface. Those who use this technique attach a small liftbag to the decompression line before the dive and wrap it right onto the reel. All they have to do is pull it free and start inflating—no wasting time tying knots.

It's true that this method takes longer to implement and keeps you on the bottom longer, absorbing nitrogen, than just tying off and ascending, but the ease and comfort of the hang justify the expenditure of time and the slightly longer decompression. Ideally, in order to compensate for unplanned difficulties, you should start your decompression procedure before it is necessary. You can always send up the line, then spend a couple of minutes in the vicinity (and within sight of the line) to use up your bottom time. Make sure when tying off that the wreck has no sharp projections that will cut the line, or that your loop will not slip off the end of a beam.

Two cautions here: If the current is stiff, the line may be pulled out of your hand before you get it tied off. When the bag reaches the surface, wrap the entire reel around a projecting piece of wreckage, then cut and tie; or pull off extra line so you have plenty of slack to work with. And when you first wrap the line onto the reel, take the time to overlap the coils neatly; you don't want the rope to tangle when it's zooming up, and jerk the reel out of your hand.

Now, with your hands free, you can monitor your gauges while timing your ascent rate. It is essential to know that the worst hazard of decompression diving is too fast an ascent. This has bent more divers than any other cause. Even if you stay at the proper stages for the specified amounts of time, too quick an ascent may cause bubble formation which further decompression will not be able to reduce. When using the *Tables,* you should also understand that too slow an ascent adds to your bottom time the difference between the prescribed ascent time to your first stop, and the time it actually took. In other words, on a 130-foot dive, if you leave the bottom at twenty minutes and don't arrive as the 10-foot stage until twenty-four minutes, your effective bottom time was twenty-two minutes. Only two minutes is allowed for the ascent. You must decompress according to the twenty-five-minute schedule: your hang has increased from four minutes to ten minutes.

Outside of resting inside a chamber, hanging below a liftbag is the most comfortable way to decompress. Since your line has so little slack, up and down movement is negligible. The bag will submerge when a high wave crest passes, then bob up in the trough. But you will not feel any sickening motion.

When your decompression is completed, you can cut the line from the reel and let it fall back down to the wreck. The natural hemp fibers will eventually decompose. Stay away from rope made from artificial materials (nylon, polypropylene). It will last a long time, and there are enough hazards in wreck diving without adding entanglement in old decompression lines.

The line should never be cut until you surface. One time I came up in a rough sea and, after making several circles and seeing nothing from the wave crests but horizon, realized that I was alone. The anchor had pulled loose and the boat had drifted away. To make matters worse, since there were divers decompressing on the anchor line, the captain could do nothing about it until they were all aboard. It was a long, lonely twenty minutes until I saw a speck on the horizon coming toward me. Had I not held onto my line I would have drifted off the wreck site in the direction of the current, not necessarily the way wind was pushing the boat, and might have been impossible to find. As it was, the liftbag acted as a marker buoy, making me easy to locate. I was not much amused when the captain jokingly shouted out as he steamed by that it was a good thing for me he had had to come back for his buoy!

Art Kirchner "ties in" the grapnel with a line, to ensure that it does not come loose during the dive.

It's a good thing this grapnel was tied in, because now only the rope holds the dive boat over the wreck.

You also might want to check that you are not too far down-current from the boat. If you think you can't make it back, someone can swim a line out to you. At least your options are open.

The only objection ever raised about this technique is: what happens if the liftbag has a leak? This question falls into the same category as: what happens if your regulator fails? While some risk exists, it can be minimized by proper gear maintenance. Like any other piece of equipment, your liftbag should be hung up periodically, filled with water, and checked over thoroughly; moreover, if you carry it on your back, where it can frequently rub up against the sharp metal of wreckage, it should be checked after every diving day. It's also a good idea, while diving, to reach around every once in a while, especially when you reach the no-decompression limit, to make sure it's still there. After a while this becomes as automatic as checking your gauges.

Whichever type of spool you choose, a personal decompression reel should be carried all the time. Even if you don't plan to use it, it should be part of your emergency or backup equipment. Then, if you accidentally overstay your time, can't find the anchor line, need to mark an area or artifact with a buoy, or just want to avoid the crowd and hang around on your own, you will have the opportunity to do so.

Since you don't have to worry about finding the anchor line, having your own reel will make you feel more at ease. You can also be more efficient in your dive: instead of spending half your time retracing your steps, you can wander freely and come up wherever you are when your planned dive time runs out.

There is just one more issue to discuss, and that is decompression computers. With the advent of space-age technology there are now on the market devices which calculate with pinpoint accuracy a diver's time and depth. Electronic movements and sensitive transducers combine with sophisticated calculators to compute the exact amount of decompression needed. Most divers see the decompression computer as a way to pick up more bottom time, since a multidepth dive is interpolated in the diver's favor. But the real benefit is the program design.

The *Navy Tables,* as helpful as they have been, are quickly becoming outdated. They were originally formulated by bending subjects in a chamber, then backing off the time on the next compression and doing a statistical analysis. The idea was to see how much of a pressure differential the subject's tissues could take without causing decompression sickness. To be sure, the *Navy Tables* have came a long way since the 1920s, when the *Attempted Salvage of the USS S-5* reported, "Four hundred and seventy-seven dives were made and only 10 percent of these resulted in a diver having the bends." This was considered to be a remarkably low percentage. They have been modified and refined since then.

Modern experimenters use an ultrasonic Doppler to detect bubble size in the bloodstream. This way, instead of allowing a nitrogen bubble to form a passable size, they allow no bubbles at all to form. This research is the basis for the slower ascent rate—twenty to thirty feet per minute—allowed by the decompression computers, and for a continuous ascent, as opposed to a staged ascent, which more closely matches the desaturation rate. Medical advances and computer science have given us safer methods and tools for decompression.

But, as with any other system, the decompression computer needs a backup. In case of battery death, component failure, flooding, or accidental damage, the diver needs an awareness of his condition. With a timepiece, the *Navy Tables,* and his ever-present knowledge, he can figure out his own decompression within allowable limits. A diver should always carry a set of *Tables,* preferably on a plastic card, so that, if he gets carried away and does not watch his time, or if the bottom is deeper than expected, he can ascend to a stage one stop below his previously calculated stop, and figure out a new decompression schedule for himself.

When all is said and done, there is no computer better than the human brain.

CHAPTER 6

Artifact Recovery

Souvenirs are a part of human nature. From the time man lived in caves he collected things: unusual-looking stones, teeth, horns, pieces of bone: all mementoes of his travels, tokens of his exploits. So much did he love these keepsakes that they were often buried with him, reminders of his life on earth which he carried into the afterworld.

Therefore, it is not difficult to understand why people today have an ardent desire to find, to buy, to possess material objects that form the substance of their lives, the sum of their accomplishments. Power and wealth are merely examples of the ways people justify their existence. What better way for a diver to chronicle his prowess and expertise than to exhibit relics he recovered himself from the depths of the sea? But before he displays his memorabilia, he has to obtain them.

Preparation

While some divers fumble around in the water halfheartedly hoping a valuable trinket will jump into their goody bags, the really serious artifact hound will start out by doing his homework. The first thing you need to know about artifacts is what they look like.

Since we are talking mainly about nautical artifacts, you need to study ships. I once saw a brass contrivance that reminded me of my grandmother's well pump. It was partially buried so, not knowing what it was, I expended very little effort on it. After listening to my description, however, one of my buddies went to the spot I mapped out and brought back a beautiful ship's whistle. The pump handle was in fact a steam inlet valve.

Libraries are full of books on shipping, both merchant marine and military. Many are replete with drawings and photographs of all kinds of vessels: from schooners to brigantines, from tankers to freighters, from tugboats to ocean liners, from submarines to battleships. Learn to distinguish one from another. Become intimate with their design and layout.

If the wreck you are diving is a steamship, you should know that the engine is always in the stern and the boilers are always in front of it. Now if you are looking for the bow, you know to swim off the boilers away from the engine.

Examine models and building plans to learn how sailing ships are rigged: deadeyes will be found on the masts, or along the hull. Since the stacks let out boiler smoke, the bridge of a steamship is usually located just in front of the first stack so the officers can avoid the smoke and soot. The bridge is where you will likely find navigational instruments and bridge controls such as the helm, the compass, and the telegraph. (When I first started diving and heard someone mention seeing the telegraph, all I could think of was a Western Union man tapping out Morse code with a copper key on a wooden block. I was unimpressed. In actuality, the ship's engine order telegraph [also called the repeater or annunciator] is a mechanical device for relaying engine direction and speed instructions between the bridge and the engine room.)

Visit maritime museums where ships' appurtenances can be scrutinized. Study them from all angles, not just as they are shown. Try to picture how they would look half buried in the sand, or overgrown with barnacles and anemones, or encrusted with coral. Ascertain what other artifacts might be found nearby. Build a relationship in your head so that, when you are exploring a wreck that is strewn all over the bottom, you can orient yourself on key pieces of wreckage.

Finally, embark on some real ships. Various marine museums keep up floating exhibits, with vestiges of the bygone days of sail docked next to pilot boats and lightships. Port cities are crowded with tankers and freighters and, even if you cannot wangle your way aboard, you can see them close up, imagine yourself swimming along the broad decks and monstrous hulls. The Navy maintains many mothballed and commemorative fighting ships, and usually provides guided tours.

This methodical approach may seem like a lot of trouble, but it will pay off in the end.

Planning

Many people, supping on abalone or lobster or speared fish, wonder why their "wreck room" shelves are empty. If nonedible trophies are your quest, you must make up your mind to select your dive sites with care. Hitting a snag for bugs (lobsters) will rarely yield much in the way of artifacts. Often, divers are reluctant to go out on boats with strangers, preferring to dive only with the club they belong to, with people they feel comfortable with and have fun with. This is great for socializing, but not always for accomplishing your goals.

Pick a charter by its destination. If you hear about a dive that is going to a wreck that has potential for yielding artifacts, then sign up for it. Stay away from barges and rock piles. Of course, you'll be putting yourself into a more competitive market—everyone else there will also be looking for artifacts—

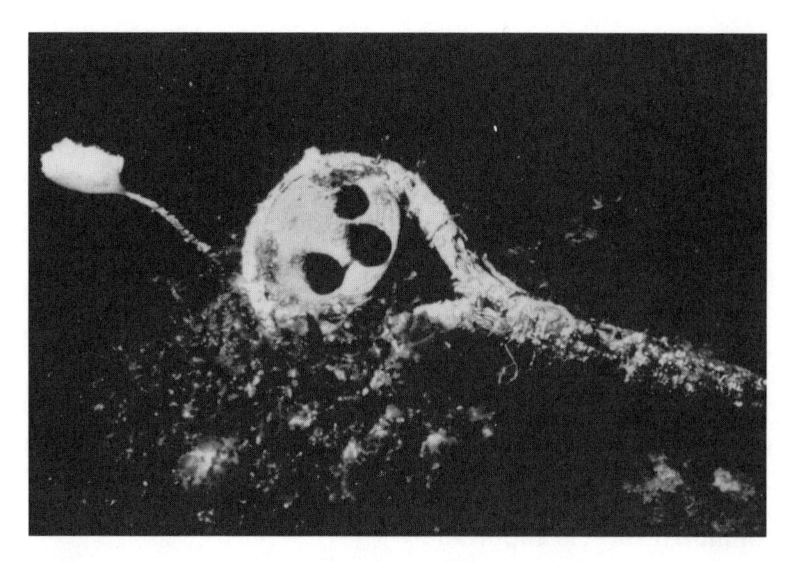

A deadeye still in its iron band lies in the kelp on a wreck near Halifax.

A deadeye, part of a sailing ship's rigging, causes a curious diver to pause.

but you are still improving your chances. You won't find oranges on apple trees.

Once you know the particular site, find out everything you can about the wreck. What has already been found on it? What location has been the most productive? What conditions should you prepare for? If it is deep, anticipate decompression; if it is intact, gear up for penetration.

Does the wreck have a name? If so, do some research about the circumstances of its sinking—if it went down fast, the crew probably did not have time to save much. What kind of cargo was it carrying? Write to the maritime museums and buy photographs of it; then go over them with a magnifying glass. Is the ship's bell on the foremast, or on a forecastle davit? Does it have a stern bridge, or is the bridge amidships? Every shred of information is important.

When your historical investigations of the ship are complete, you have optimized the odds by knowing what to look for, and where to look for it. Now let's discuss *how* to look.

Locating Artifacts

Since most artifacts are not found lying loose and in the open waiting for a diver to pick them up, before we enter the water it is wise to mention tools. Every occupation, from office work to auto mechanics, has its specialized implements. Artifact retrieval has its own, adapted from topside trades. While some divers go down looking like hardware store representatives, there are a few basics that will suffice for most jobs.

Understand that any hand or mechanical tool that can be used in your garage or workshop will work underwater. The major disadvantage of the wet environment is loss of leverage. That is why the crowbar is the most valuable instrument you can have. It is cheap and expendable. If Teddy Roosevelt had been a wreck diver, he would have said, "Dive softly, but carry a big crowbar."

Although it is heavy and awkward to carry, the best all-around size is between eighteen and twenty-four inches. I

On the *Ioannis P. Goulandris,* a Broco torch goes to work on the five-inch steel shaft holding the helm onto the steering gear.

The helm from the conning tower of the *Tarpon.*

keep mine on the back of my doubles, held there by one thick rubber band (actually an inner tube slice) from the manifold and around the hook, and one around the base of the tanks. It is out of the way, and with some practice you learn how to reach behind you with one hand, hook your thumb over the bottom band, pull it off, then jiggle the crowbar out of the top band. Of course, your buddy can do it quicker. Anyway, most dives are spent looking for artifacts, not working on them. This way, at least, you will always have it with you, but you won't have to break your arm carrying it.

The only other tools you need to lug around all the time are a hammer and a chisel. An eight-inch chisel can live in your goody bag, but a three-pound maul will destroy your wrist. Attached to your weight belt, the hammer becomes part of your buoyancy control. Either slip it through the belt loop outside one of the weights, or put a D ring on your belt and a screw hook on the business end of the hammer, then clip it in place. (Do *not* put snap hooks on your weight belt because fishing line and cables can snap into them accidentally.)

You might also want to carry a screwdriver. Whenever possible, use finesse. Since you might inadvertently damage an artifact, employ wrecking tools only when the more delicate operations fail. Unscrewing a gauge from a bulkhead is preferable to prying it off and possibly curling the lip or bending it in half. You are not knocking down a building; you are removing something of great value—something with which you want to decorate your house.

Only minor maintenance is required with these solid steel items, but if you take down such things as pliers, vise grips, or a pneumatic chisel, using copious amounts of a rust inhibitor like WD-40 after every dive will keep them in working condition. If you use these tools often, keep a bath going at home, and simply soak them when not in use.

The major difficulty in finding artifacts is in recognizing them, especially when they are heavily encrusted. The trick here is to look for artificial shapes: straight lines, geometric curves, depressions, and sudden changes in elevation. A problem in northern wrecks is darkness and the monotone color;

Sometimes a porthole requires extensive cleaning underwater before you can figure out how to remove it. Working in depths of 180 feet, as in this case, can severely limit your effective time.

On the light cruiser *Wilkes-Barre* Jeff Pagano carries a porthole he removed from the superstructure.

in southern wrecks, coral encrustation. Only experience will overcome these visual obstacles.

Concentration is important. Instead of just looking, you must perceive. This is hard to explain until you have swum over an area ahead of your buddy, and he picks up the artifact. When you scratch your head and ask yourself, "Why didn't I see that?" you will know what I am talking about. Do not just play your light glancingly over an area and move on—stare at it, study it, observe it. Hold your light at an angle so it creates shadows on an otherwise flat surface.

Do not become distracted by lobsters or other marine creatures. Watch the fish some other time. Also, do not become blinded by your expectations—some divers want a porthole so badly they fail to notice something better lying right under it. If you see one artifact lying half exposed in the sand, stop and study the area before you muck up the visibility by ripping it out. More things may be lurking half hidden, only to become fully obscured by your impatient lunge.

If you find one artifact, scour the area. The protruding rim of a dinner plate may mean that silverware is buried nearby. One gauge on a bulkhead indicates that others should be mounted at similar heights. Deadeyes come in rows: look to the left and right. Once I found an empty porthole on a huge section of hull plate lying flat on the sand. Instead of being discouraged, I followed the bulkhead, counting the number of partitions until I came to the next empty porthole. I did this until I reached a spot where a piece of steel covered the place where I thought the next porthole should be. When I lifted the steel off, there was the porthole, untouched. A swift blow with a driftpin punch and the glass and frame were detached from the cast-iron backing plate.

While I am on the subject of portholes, there are a couple of things you should know. One is that portholes open from the inside, so in order to remove the hinge, on an intact wreck, you must be prepared for penetration. The other is that all those round holes you see so often in blown-up wrecks, unless they are surrounded by bolt holes, are weight-reducing aper-

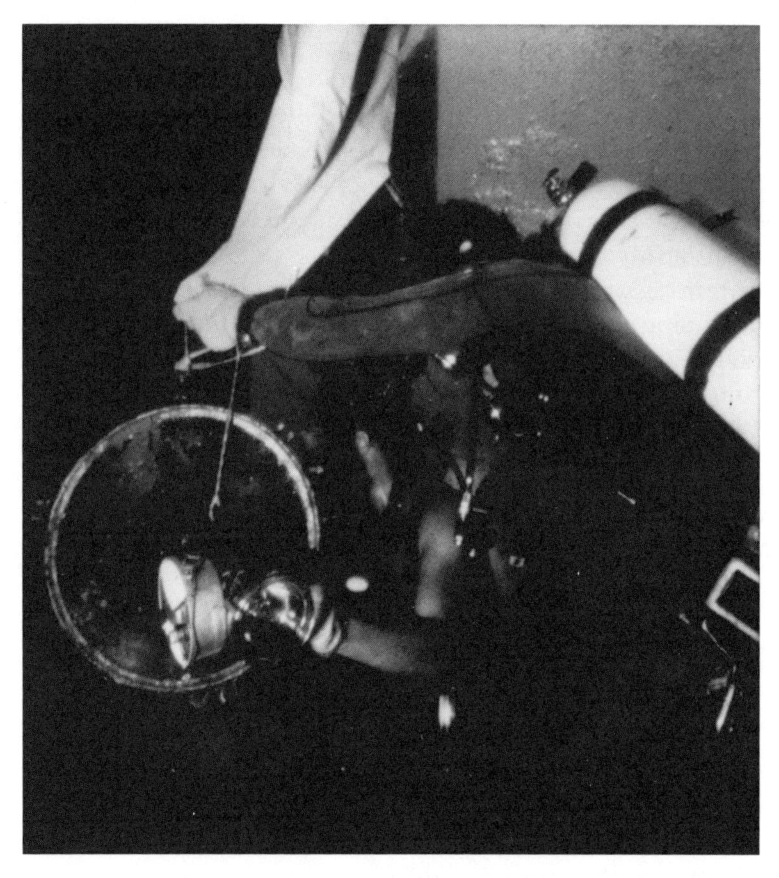

Divers Rick Jaszyn and Jeff Pagano attach a liftbag to a porthole, a souvenir from the *Wilkes-Barre.*

tures in the double-hull support struts. Beyond that, the knack of removing portholes is an art—and an article—in itself.

Now let's turn to sand-searching techniques. We can't all be equipped with a prop wash, or work wrecks in shallow enough water for one to operate properly. But a comparatively inexpensive operation for those of you with your own boats is an airlift. This requires little more than a medium-sized compressor and a long fire hose with a five-foot length of four-inch pipe on the end. The fire hose feeds into a curved fitting so the air is pumped out one end of the pipe, causing suction at the other end. This is a double-edged sword, since you can either suck sand up or blow it away. I have used an airlift at a depth of over a hundred feet, but the depth at which an airlift will function depends on the capacity of the compressor.

Even with hand implements, however, you can move quite a bit of sand. Besides just fanning it with your hand, you can pick up a large clamshell, or take with you a flat board or piece of plastic. Ping-pong paddles are great. You'll be surprised how quickly you can dig a sizable hole. I often carry a three-pronged garden cultivator, the kind that is about a foot long and used for weeding flower beds. With it I scrape through the top layer of sand where I expect to find artifacts. When diving for bottles in mud or heavy silt, just jab your dive knife downward and listen for the peculiar clinking sounds; then reach in with a gloved hand (for protection against broken glass) and pull out your find. It is amazing how many artifacts are missed because of a fraction of an inch of sand or mud.

The Art of Lifting

Put away the bruiser tools and let's get your precious find on the boat. The first rule is: never carry anything in your hand. You are too likely to drop it. If it is tiny, like a coin, slip it inside your glove. Anything larger should go into a mesh goody bag.

Place your tools in the bottom of the bag, and your artifacts on top of them. If you are bringing up bottles or glassware, you need to be extra cautious. They might get slammed together during a surface swim in rough seas, or banged

The liftbag attached to this *Proteus* window is also tethered to a decompression line of quarter-inch sisal. Tying the line to the wreck will prevent the liftbag from floating away with the current. The line is biodegradable.

against the side of the boat while being lifted over the gunwale. If you expect to find such items, take some cloth wrappings with you. Otherwise, use common sense—return with the items one at a time, take more than one goody bag and stagger them during the haul, use your buddy's bag, or hand one item to a passing diver.

Anything too large for a goody bag, or too heavy to raise safely, needs a liftbag. These large, heavy-duty plastic balloons are invaluable for light salvage. They come in a variety of sizes, with lifting capacities of from fifty pounds to several thousand. I always carry a two-fifty, with a hundred-pounder for a backup. When attaching the snap hook from the liftbag lanyard to your goody bag, never clip it to the metal handle. This is spot-welded and might break off from excessive weight. Instead, run a quarter-inch nylon line through the mesh and around one of the hoops, and lash the liftbag here. Do the same with a plastic-hooped goody bag.

If you want to fasten the liftbag directly to the artifact, you will need a lifting strap that will not be chafed through by abrasion against jagged metal. The handiest is a stainless steel cable with an eye cinched at each end, available at marinas and boating stores. Wrap it around the artifact, then choke it by reeving one eye through the other. As soon as the tension is taken up, the weight of the item tightens the loop. The problem with cable is that it is practically impossible to grip when pulling a weight of over fifty pounds out of the water; and, as it wears, sharp strands break off and cause dangerous and painful puncture wounds in the fingers and hands. The advantage, for smaller items, is that the eye will fit through the bolt hole of a porthole backing plate.

Better than cable is a rubber-faced nylon strap. These come in two- or four-foot lengths, and sport different-sized stainless rings at either end so one can be pulled through the other like a choker. Snap on the liftbag, and use the strap as a lifting line on the surface. The 4,000-pound breaking strength is more than adequate.

Now that the liftbag is rigged and ready to go, you have to fill it with air. Ideally, in order to save air, you should get your

The *Manuela*'s bell, after it was exposed. Notice the clapper inside. The thick, dead encrustation shows that the bell must have been above the seabed for a long time before it fell into the sand and became buried.

buddy to do the job. Otherwise, you will have to hold your breath, take the regulator out of your mouth, stick it under the opening and push the purge. Make sure you spread the plastic, or your precious air will slip unnoticed up the outside of the bag. This operation can get tiring as you alternately take a few puffs and dispense air to the bag.

To alleviate the strain you can use your pony bottle regulator, or breathe off it and use the main for filling. A problem can arise in the latter case when you forget to switch your regulators back and you drain your pony bottle. If your regulators are not well maintained, they may start freeflowing. A much more serious hazard occurs when the exhaust tee or the hose gets snagged in the liftbag shrouds and the artifact finally takes off, dragging you with it.

A simple way to avoid these problems is to use a liftbag inflator. This is a simple but handy contrivance that snaps onto a BC hose from a low-pressure port. I carry one all the time, using tie-wraps to keep most of its length joined to my submersible pressure gauge hose. Now you can breathe and fill at the same time.

You would be surprised at how many divers swim away after watching their hard-earned artifacts leave the bottom as if the liftbag will magically deposit its load on the boat. These are the people who never swim after anyone else's bag, but get ironically furious when someone does not chase theirs. Your liftbag is your responsibility.

If you think you might be sending up a liftbag, make arrangements to have a lookout dressed and ready to dive in after it with a safety line. Even then, with a stiff current running, it might get away. With a compass bearing you can pursue it with the boat—but not until all the divers are aboard. And in a choppy sea, every cresting wave looks like a liftbag. The solution is to tie on your own safety line while you are on the bottom.

Using your decompression line, or a lather's tie-wire reel, fasten a rope to the artifact—not the liftbag—before sending it up. As the liftbag starts its ascent, hold onto the reel and let the line spin off. After the liftbag hits the surface, but before you

Diver controlling the ascent of his liftbag as he guides it up the anchor line.

Art Kirchner ties on a safety line before hauling an artifact out of the water.

cut your end of the line, tie the rope to a convenient piece of wreckage. This way, the surface current will not rip the rope out of your hands before you get the knot tied, and your artifact will dangle below the liftbag, securely attached to the wreck.

Should the liftbag dump or sink, or the shrouds tear away, you can always return to the place where you tied off, follow the rope to the artifact, correct the problem, and send it up again. Meanwhile, you can continue your dive. Afterwards, you can swim out another safety line from the boat, fasten it, cut the bottom line, and have the liftbag towed back. Alternatively, you can come up the safety line, even do your decompression, then, if the bag appears sound and there is no surface current, cut the bottom line long and use it as a tow rope.

Another method is to partially inflate the liftbag and tow it on the bottom back to the anchor line. You can either swim it up, or clip it to the line with a carabiner and let it ride up to the bow of the boat. The problem with the latter case is that it might collide with decompressing divers as the liftbag, air expanding, soars to the surface. But experienced wreck divers will understand.

So that you do not get the captain upset by scratching the paint or gouging the wood when a large artifact is hefted aboard, take along a collision mat or a sturdy canvas tarp to lay across the gunwale. And just in case the artifact is dropped, tie on a thick safety line, or attach a buoy with enough string to reach the bottom.

This is still not enough. I have seen too many artifacts reach this stage only to be left on the deck to get stepped on or have tanks dropped on them. Nor is it desirable to have a 200-pound helm stand rolling into divers and their gear; lash it down so it cannot move.

Anything small should be stowed carefully, even if it means throwing out your food and using the cooler. It may be a rough ride home, and a china teacup that survived a traumatic capsizing and a hundred years under the sea may not make it intact to that yearning mantelpiece. And what kind of story would that make?

CHAPTER 7

Artifact Preservation and Restoration

While many nautical artifacts found in homes and stores have never known the destructive environment of the sea, those items recovered from underwater—especially from saltwater environments—may be in a condition undesirable to the owner, or unstable, when removed from their prolonged liquid bath. For this reason we have to differentiate between "restoration" and "preservation," and define our terms.

Restoration, like beauty, is in the eye of the beholder. To some, highly polished brass that gleams in the light with a golden luster is the ideal to be sought in restoring that metal. To others, the green patina that is the result of oxidation of the copper content lends more of an air of the sea, evoking the brine-filled atmosphere and maritime conditions which prevail during a long ocean voyage. Be that as it may, restoration can mean either of these extremes, or any grade in between.

Preservation, on the other hand, is not so arbitrary. Without the proper techniques, some materials are radically affected by saltwater immersion and must be brought back gradually to a stabilized condition. Otherwise, wood will shrink and split, leather will twist and wrinkle, and iron will crumble into an unrecognizable pile of junk and dust.

This chapter will cover each material separately, will discuss the "backyard" treatments available to the individual, and will describe the tools, chemicals, and safety measures necessary to bring one's nautical antiques to the desired state which will not only enhance their appeal but increase their value.

Water, although described chemically as a universal solvent, can also act as a universal preservative. Whenever artifacts of any material are recovered from the sea, they should be stored immediately in fresh water until the necessary treatments can be administered. Tap water will do, but when dealing with ancient or more delicate relics, such as organics, distilled water, without chemical additives, offers a distinct advantage. This way you can put off indefinitely any work processes until you gather the materials and have the time to do the job, without further damaging the objects to be worked on.

Brass and Bronze

By way of definition, brass is an alloy of copper and zinc. It has a yellowish tint. Bronze is more reddish in color and is basically an alloy of copper and tin, with other metallic substances, especially zinc, added in small quantities. The relative amounts of the components are variable.

Brass and bronze are highly resistant to corrosion, which makes them ideal for ship furnishings and working parts. The green patina often seen—the bane of many a sailor—is a thin layer of natural oxidation, called verdigris, which does not damage the underlying molecular structure. Brass and bronze, therefore, do not need to be preserved when recovered from the sea, but merely restored to a condition deemed suitable to

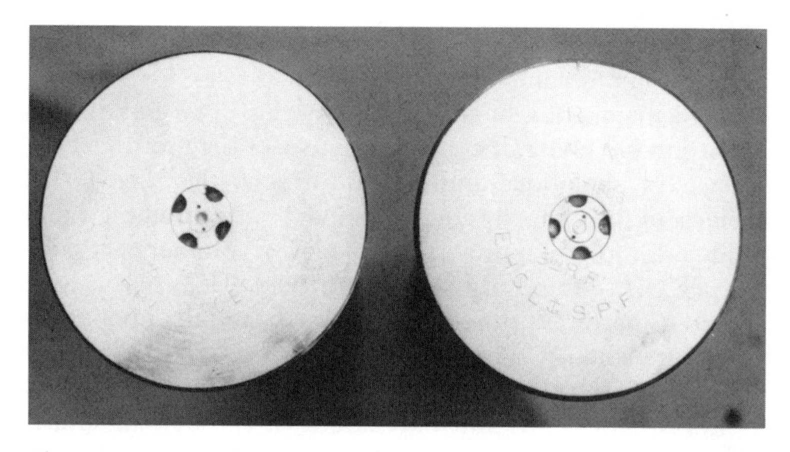

The bases of two three-inch gunshells, showing the size and dates of manufacture.

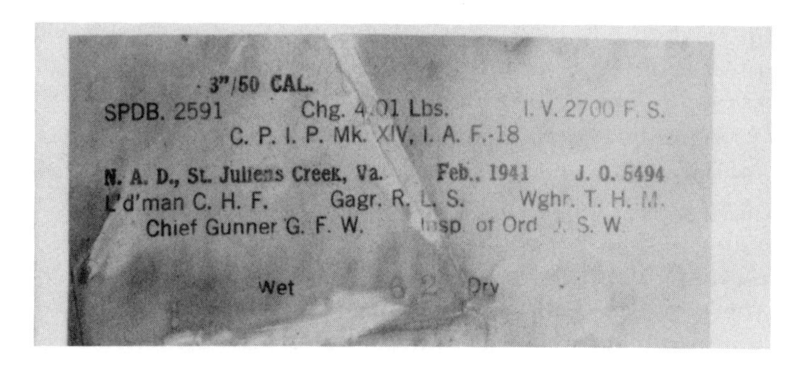

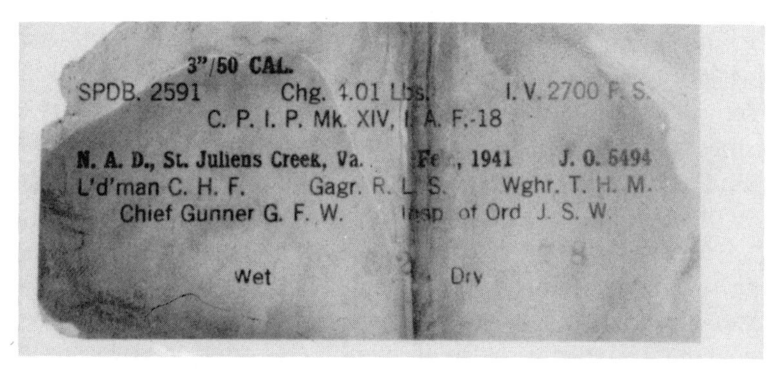

Packing slips taken out of the three-inch gun shells found on the *Kolkhosnik.*

the collector. (From here on I will use the word brass as a collective noun to stand for bronze as well.)

The major consequence of submerging brass in the ocean environment is that it collects a veneer of sea growth: coral, barnacles, anemones, and various forms of plant life. To rid objects of these smelly encrustations requires nothing more than a slow dip in muriatic acid. This is a commercial-grade, 20 percent solution of hydrochloric acid (HCl) and can be found in hardware stores or wherever swimming-pool accessories are sold—it is used as a pH adjuster. It comes in one-gallon jugs. If larger quantities are needed, locate a chemical outlet or brick yard and purchase it in carboys containing about twenty gallons: the cost per gallon will be about a quarter as much.

Caution: Muriatic acid is a strong corrosive, and dangerous. When handling, wear rubber gloves and *always* use eye protection: a plastic face shield, or goggles, or both. Also be careful not to inhale the fumes.

A plastic container, preferably with a lid, should be placed outdoors in a well-ventilated area that can be cordoned off from children. Pour the acid into the container and add an equal or double amount of water: this not only cuts down the strength of the acid while stretching its usage, but is easier on the brass object. Gently place the artifact into the acid, holding your breath. The acid will start effervescing immediately. If the object is not covered fully, add more water. To prevent fumes from escaping, clamp down the lid, or cover and seal with a plastic trash bag. (*Note:* Those of you who had high-school chemistry will remember that, in order to avoid a possibly violent reaction, water is never added to acid. Instead, acid is added to water. However, muriatic acid is dilute enough to obviate this caution.)

Most marine growth will be dissolved or swept off in a day or two. Hard corals will require more time. It is better to use more time rather than stronger acid in order to prevent the surface from pitting. Also, if the object is not fully covered the fumes will create a green line where the liquid ends. This will be impossible to remove, even with further soakings.

When the process is completed, the acid may be stored in plastic jugs and used again. As long as it bubbles when encrusted artifacts are put in it, it is still good. When it is exhausted, either spike it by adding more acid, or dispose of it by pouring it down an outside drain and flushing it with copious amounts of water.

A while back, one of my dive buddies called in a state of near panic. He had acid-dipped a porthole, polished it, and hung it on his living room wall—all in a span of a few days. Within a week it had turned completely green! An unsightly and pungent powder had formed on the surface, and this was gradually flaking off onto his new carpet.

The chlorine from the muriatic acid was leaching through the outer corrosion layers and forming chlorides. The same will happen, to a lesser extent, to any piece of brass taken out of seawater and left untreated, as chlorides will form from the salt content—remember that sea salt is sodium chloride. As if it were a medical condition, a slow greening process known as bronze disease can occur months or years later, and can be difficult to cure.

For this reason, all brass objects, whether acid-dipped or not, should be soaked in fresh water for at least a month; longer if the brass is exceptionally porous, or exceedingly old. The water should be changed about once a week. A box of baking soda dumped into the tub can further neutralize any remaining acid, natural or otherwise. This base, or alkali, can prevent bronze disease, but to reverse an ongoing case the only treatment that might succeed is soaking in ethyl alcohol (pure grain alcohol).

As an intermediate stage for ease in polishing, or for objects too large to soak in acid, sandblasting is recommended. A small sandblasting unit can be picked up at any hardware store or home center, complete with bucket, hoses, and a variety of nozzles. However, a compressor is necessary to operate it—one of small capacity is a moderate-sized purchase. A fine sand such as double-aught banding sand, or glass bead with the consistency of talcum powder, must be used. A larger-grain sand, such as beach sand, will seriously pit any smooth surface and should be avoided. By using a tent, or blasting against a wall on a thick tarp, most of the sand can be reclaimed. It then needs to be poured through a screen, to filter out large flakes that have been blasted off. This will prevent clogging the unit and damaging the artifact.

Now we are ready for polishing. This can be done with either double-aught steel wool (Brillo pads are great) and gobs of elbow grease, or a fine wire wheel on a reversible drill. Why reversible? After a few minutes of polishing the bristles get bent backward so the cutting edge is dulled. Reversing the rotation increases the abrasiveness by pointing the tips of the bristles forward. Never use a coarse, or even a medium-grade, wire wheel.

You can either leave the relic here, or go one of several ways. For a goldlike, highly reflective surface, apply buffing compound and burnish it in with a soft wheel. You will end up with a museum piece. For that "at sea" look, a short soak in vinegar will give a slight greenish tinge.

In either case, you might want to add several thin coats (rather than one thick coat) of clear acrylic, best applied from a spray can. This will seal the finish with an invisible veneer that will prevent retarnishing, and allow handling without leaving grease prints or smudges which are almost impossible to remove once embedded.

Please note that if the freshwater soaking was not long enough the acrylic will not prevent the artifact from turning green. This is not an effect of exterior oxidation, but of chlorides leaching out from within. The acrylic can always be removed if additional work or treatment is necessary.

Lead, tin, and babbitt metal can be treated in the same manner.

Gold, Silver, Copper, and Pewter

These metals are lumped together not because of any similarity in composition (the first three are elements, while pewter is tin alloyed with lead, brass, or copper) but because the same treatments can be used for all of them. However, the optimum cleaning method will be given for each metal.

Gold is almost always found gleaming like the day it was minted. The noblest of the metals, it is so resistant that it will not react with either organic or inorganic compounds—marine growth cannot get a foot on it, so to speak. Occasionally, it may be found stained, or tinged, by other metals in the vicinity. Since gold can only be dissolved by aqua regia (a mixture of hydrochloric and nitric acids) almost anything can be used to dissolve these slight discolorations. Gold, however, is also one of the softest metals: it should never be rubbed or scraped with anything more abrasive than a soft cloth, or it will scratch.

Silver, when found in chunks such as bullion or conglomerates of coins, is fairly resistant to the ravages of the sea. Exposed surfaces, unless "saved" from electrolysis by the sacrifice of nearby metals, will be in poor condition and cannot be restored. Marine growth can be removed by the muriatic acid treatment, although in this case it is better to soak the silver in household vinegar. The solution will effervesce slowly, and you can actually watch the particles break away and float up to the surface. Once loosened, rub the silver object with a nylon pad such as that used for cleaning Teflon surfaces. Afterwards, soak in fresh water for a couple of weeks to leach out the acidic residue, and brighten with silver polish. On the other hand, if the silver is merely tarnished with black silver sulfide, try rubbing with a paste of water and baking soda.

Copper, unalloyed, is a fairly weak metal which dissolves all too readily in the sea. Natural electrolysis causes a mild electric current which turns atoms into ions, thus reducing their cohesiveness. They then float away, leaving "holes" in the material. This process cannot be reversed. Strong acid treatments will completely dissolve delicate copper sheet, and should be avoided. Vinegar is the only viable method of cleaning other than electrolytic treatment.

For electrolytic treatment, which is a complicated procedure, you will need a plastic tub (large enough for the object) filled with water and a 5 percent solution (by volume) of lye, a car battery or battery charger to generate direct current, and a sheet of scrap iron. Using copper wires, connect the object to be cleaned to the negative pole and the iron to the positive pole. Submerge both in the solution, but not touching each other. An electric current passes from copper (cathode) to iron (anode), ripping loose chlorides from the artifact and depositing them on the scrap iron. Clinging marine growth is jarred free and falls to the bottom of the bath, or is dissolved away. Length of soaking depends on the size of the object and the amount of growth, ranging from hours for coins to weeks or months for something large like a cannon. The main advantage to this method is that the copper is plated in the process, making it more chemically stable. This galvanic reduction can be used for any copper alloy, as well as silver and pewter.

Some tips for treatment. The current should be turned on before immersion and turned off after removal of all artifacts, to ensure against plating by undesirable corrosion products. If you want to get fancy, a variable-speed auto transformer will allow you to control the amount of current. Too high an amperage will cause bubbling of the solution and can damage the object by reacting too fast. A full twelve volts may be fine for a large item, but too much for a coin. And turn over your relics periodically so surfaces touching the container have access to the solution.

Pewter on its own is a fairly stable compound. But when leaning against iron, such as a ship's hull, the contact of dissimilar metals causes electrolysis in which the pewter is sacrificed. It may appear deeply corroded, or pitted, a condition that cannot be corrected. Pewter, too, is often plated with silver, which, because of its thinness, usually dissolves unless the object is covered with silt or sand. In either case, the treatment is a delicate one.

The object should be soaked in water for at least a month, then dried. Now use a simple technique called "flecking." Gently tap the encrustation with a hard tool, such as a screw-

A dessert plate from the *Andrea Doria* showing oriental design.

driver, to chip off particles. So the soft metal is not dented, use the rounded shaft of a Phillips screwdriver rather than the squared edges of a regular shaft. Also, do not gouge, as with a chisel. Once most of the hard stuff is off, use a wooden pencil to scrape the surface and get into the grooves. A rubber eraser can be used to polish.

Should this method fail, you can revert to the vinegar treatment. This is fine for pewter alone, but if there is still silver plate clinging to the object it will more than likely peel off. Finally, a buffing wheel can restore a high sheen, as can pewter polish. Clear acrylic sprays can be used on pewter, as well as gold, silver, and copper.

Earthenware, China, Porcelain, and Glass

Ceramics have been made by mankind since the days of antiquity. Basically, a ceramic material is clay, or a mixture of mud and water, which is either dried in the sun or baked in a kiln until it has solidified. It has a rough and uneven surface, although sometimes it is "glazed" on the outside, a process in which a material called frit (finely ground glass, clay, and water) is applied by dipping, spraying, or brushing. After drying, the ware is fired and the coating becomes smooth.

Earthenware, although fired, is generally unglazed. China can have a wide variety of minerals mixed with clay, every manufacturer's formula being different. It is glazed to give it a smooth, shiny appearance. Porcelain, known as vitreous china, is fired at a much higher temperature, which, because of the greater fuel cost, makes it more expensive. It also makes it waterproof.

Glass is merely silicon dioxide (sand) that has been melted, then molded during cooling. Adding minerals to the molten solution gives it color. Of all the categories of marine artifacts, glass objects are the easiest to clean. Within the time span of naval history, glass and earthenware products have been impervious to salt water, microorganisms, and all forms of marine plants and animals. Although coral and barnacles may grow on them, other than an occasional etching of glass by the secretion of natural glue, the main hazard, as in the kitchen, is breakage.

Heavily encrusted artifacts should be left encrusted in order to lend an air of authenticity. A barnacle or a clump of coral on a glass or dinner plate substantiates the underwater environment from which the object was removed. You can work around the encrustation by cleaning adjacent surfaces with soap and warm water, washing with a cloth, or rubbing with a nylon pad.

Any concretions you want removed should be worked on while wet, when they are soft. The use of small picks or dental tools is time-consuming but safe, and allows selective removal. For gross removal, soaking the object in a water-softening agent, such as Calgon (diluted to 10 percent), will do the job.

Orange-colored stains, the result of iron oxides from rusting hull plates, can be removed with an application of vinegar or a mild abrasive such as cleanser. Earthenware, porcelain, china, and glass artifacts can be dipped in muriatic acid without harm—*but* be careful of china with gold leaf.

Unlike the inks used to implant insignia, shipping line crests, and manufacturer labels, all of which are applied before glazing, gold leaf is put on after this protective process. There is not much material in a leaf of gold which has been pounded until it is only .00001 inch thick. Long submersion in salt water, with its varying pH levels and strong ionic potential, will loosen the adhesive used to attach the gold leaf. Frequently, the gold leaf will wash off under nothing more than faucet water pressure, leaving only the underlying embossing template to show where it had once been.

If gold leaf is present, cleaning can only be done carefully, by hand. A cotton swab dipped in vinegar can be used to scrub close to the gold leaf, while a sponge can be used on the rest of the piece. The gold leaf is usually too delicate to be touched—and should be left alone.

The final step necessary for restoration and preservation is the freshwater treatment. Even though the glaze on china and porcelain appears smooth, the surface is in fact very porous when viewed under a microscope. Sea salt permeates the material and, in the absence of water, will crystallize out, forming visible lumps of pure, and digestible, salt. As ice can cause rocks to split, salt is hard enough to crack or exfoliate the glaze. The surface will splinter until it resembles a spider web, a condition known as "checking." Soaking in fresh water for at least a month with weekly water changes will force the sodium and chlorine to stay ionized until flushed out.

Iron and Steel

At first glance, one might wonder why anyone would bother to recover ships' parts made of iron and its carbon alloy counterpart, steel. Rusting hull plates have very little intrinsic value other than as scrap. But many interesting finds, from anchors to cannonballs, are in sharp demand by collectors. The above-mentioned artifacts serve admirably as lawn ornaments and paperweights, and can effectively blend antiquity with usefulness.

All iron products rust more rapidly when removed from the sea. As the metal dries, air enters the pores, and the presence of so much free oxygen accelerates oxidation—the cause of rust. This unwanted iron oxide forms flakes, and will chip off until nothing is left but a handful of dust.

Here, instead of acid, we must use a basic solution not only to rid the iron of marine growth, but to prevent disintegration. The chemical needed is sodium hydroxide, also called caustic soda, or lye, obtainable at your neighborhood supermarket in the section with drain cleaners. Prolonged soaking in a 5 percent solution (by volume) of lye mixed with tap water starts the preservation process. For fairly new ironmongery this phase can be completed in a couple of months, but for real antiquities such as cannonballs, old rifles, or harquebuses, upwards of a year is necessary.

A plastic container must be used, not because the lye would dissolve a metal pail, but because the ion-hungry iron object would strip the galvanized coating off the bucket, and the liquid would then escape through thousands of tiny pinholes. There will be no effervescing, since marine growth is not readily dissolved, but only loosened at the iron-organic interface. However, the bath should be tightly sealed to prevent people—especially children—from putting their hands into it: it looks just like water. Lye is just as reactive as acid, but on the other side of the pH scale.

After the long soaking, any lingering encrustation must be chipped or scraped off. Now we will accede to the iron's demands and galvanize it. This is nothing more than the addition of zinc, a metal which prevents electrolysis of iron by having more ionic potential. Boat owners know that zinc blocks must be attached to the hulls of their craft near the propeller shaft and rudder for this purpose. Buy the zinc at a boat yard or scrap metal dealer.

Hold the blocks about six feet in the air, over the lye solution, and melt them with a propane torch. This allows the droplets to cool slightly before hitting the liquid, where they will spatter and form oddly shaped pieces of shrapnel with a large surface area per volume. When the zinc is cool, put on rubber gloves and pack the zinc shards around the iron so the object is completely covered. Let it sit for a couple of months until the galvanization process is complete.

Remove the artifact and let it dry. Now you can either leave it alone, apply several coats of clear acrylic, or paint it black with a rust-proofing spray paint, as desired.

For something as large as an anchor, unless you have a swimming pool you do not mind converting into a preserving bath, you will have to forgo a treatment process. But, because the object is so thick, there will usually be plenty of solid material left once the gross concretions and surface flakes are chipped, scraped, or sandblasted off. In this case, make sure you do a good job of sealing with an acrylic or flat black rust-inhibiting paint in order to prevent further oxidation. If the object is to be left outside in the elements, where weathering and natural sandblasting will present a problem, be sure to reseal it at least once a year—more often if you live at the shore. In this case, polyurethane, applied thickly, is your best bet, if you don't mind the glossy look.

Wood, Leather, and Paper

These are all organic substances in that they were once parts of living plants and animals. Being animal and vegetable, rather than mineral, they are susceptible to the same problems encountered by all natural fabrics and boards—being eaten. They are, in essence, biodegradable.

Besides cellulose-reducing bacteria, the aquatic teredo is the most prevalent and most voracious destroyer of organic substances. Contrary to popular belief, this tiny animal is not a worm, but a wood-boring mollusk. Unless wood is covered by sand or silt it is soon riddled with holes as these little creatures literally eat out their homes.

Barring these hazards to survival, over which we have no control, the real problem with wood and its affiliate substances is shrinkage. The living matter from which these artifacts are made is composed of cells, each one a microscopic cube with six flexible walls. When submerged for long periods of time, wood swells as water enters the cell structure by osmosis. The resultant internal pressure pushes outward on the cell walls, bowing them. Each little bow, multiplied millions of times, becomes visible as swelling. After the wood dries, the cell walls, which have been stretched, lose their elasticity, causing each cell to shrink to less than its original size. Cohesion between the cells must give, causing the object to crack from the undue stress. It is this shrinkage we must control.

The first call of order is not to allow the object to dry out. After recovery, keep it wet either by submersion or, where this is not possible, by draping it with a wet towel and sprinkling water on it until you can get it reimmersed; alternatively, you can wrap the item in plastic—the condensation will suffice to keep it damp enough until you get it home. As soon as possible, start soaking the object in a freshwater bath to flush out the salts and minerals. According to the age of the wood and the time it has spent underwater, this phase could take anywhere from one month to six. I like to place small objects in

This lump on the wall is a disguised boiler room gauge from the *San Diego*.

This gauge from the *San Diego* (christened the *California* in 1907) shows how a little time and care can turn an unidentifiable lump into a beautiful artifact.

the toilet tank, where there is a constantly changing supply of good, clean water.

If there is unwanted marine growth, such as a coating of coral, it can be removed by a muriatic acid dip—the acid will not harm the wood. But if there is attached iron, such as the band around a deadeye, this method cannot be applied since the acid will dissolve the metal. Also, it is best to do the acid treatment in the beginning so the fresh water can flush out the chlorine.

For actual preservation, however, we need to imbue the cell structure with a chemical strengthener. Acetone works well on newer wood, and ethylene glycol (antifreeze) can be used in a pinch although the coloring agent may tint the final product. Ethylene glycol is a gasoline byproduct, chemical formula $OH–CH_2–CH_2–OH$. This is the simplest polyhydric (many hydrogens) alcohol, with a molecular weight of 63.

Better than this is *poly*ethylene glycol, hereafter referred to as PEG. This microcrystalline wax is a chain of ethylene glycol molecules linked together into any desirable length, or molecular weight, depending on the intended usage. For our purposes, 4,000 molecular weight is the best, although the more readily available 3,350 is almost as good. Check your local commercial chemical outlet for availability and price: be forewarned, it is expensive. PEG comes in either liquid or powder form. If it comes as a powder, it must be mixed with water according to instructions.

PEG works by impregnating the cell walls and forcing out the water. In addition to filling the voids, the chains intertwine like billions of miniature ballpoint pen springs that keep the wood swollen. It also binds the cellulose to the lignin, thus adding structural stability to the weakened members. Leave the bath uncovered so the water can evaporate. This concentrates the PEG and forces more of it into the cells.

The object should be soaked for a minimum of six months—longer if the wood is very old. Archeologists in England have soaked thick oaken planks of Roman vintage for as long as two years in order to be assured that the PEG penetrates all the way to the interior cells—and for some woods they say that as long as ten years may be necessary. They also recommend temperature and humidity control: particularly fragile items should be maintained at room temperature (don't leave something soaking in your unheated garage during the winter, where the solution might freeze solid) and 100 percent relative humidity. They even go so far as to start the procedure with PEG 1,000 in order to quickly reach the inner core, then go to a PEG 4,000 to seal it in. But this is only done for archeologically ancient woods.

This method, the best there is, is not perfect. There will still be some shrinkage, perhaps as much as 10 percent, but usually not enough to cause serious damage. In the case of coral-encrusted wood, this shrinkage can be advantageous. The coral will not shrink as the wood drops away from it, and eventually it will crack and flake off.

After freshwater rinsing, some PEG might form on the surface. Pour hot water over the object to remove unwanted wax.

Now, more than a year later, the wood can be left natural, or it can be coated. Varnish and shellac leave a store-bought glossy look which, if you like it, is fine. Otherwise, use varnish or shellac only when necessary to hold crumbling parts together. Sprayed polyurethane will give a natural finish while providing protection as well as a sealant. The wood can be sanded and rubbed with linseed oil to produce a nice, waxy finish reminiscent of seafaring days.

Follow these same procedures for leather and paper, taking care to keep paper flattened between two sheets of glass to prevent it from curling. Because of the thinness of paper, a much shorter treatment time— a month or two—may be used.

Bone and Ivory

Bone and ivory are only partially organic. Extended submersion dissolves collagen and fats, leaving the remaining material porous and brittle. Upon drying, bone and ivory items are subject to shrinking, cracking, and warping. But unlike wood, bone and ivory consist of some 60 percent calcium compounds. This poses problems in both cleaning and preservation.

Since marine encrustations are also composed of calcium compounds, any chemical which breaks them up will destroy the bone and ivory as well. Mechanical means must be used to rid the objects of unwanted concretions. Vinegar can be used, but only for an hour or two at a time. Then the objects must be rinsed thoroughly in fresh or distilled water. This process can be repeated as often as necessary, as long as you watch the artifact closely: if any deterioration occurs, stop at once. A safer but much longer treatment is a 10 percent solution (by weight) of sodium hexametaphosphate (Calgon); still, there is no guarantee that the encrustations will be sacrificed before the artifact. You may just have to leave some concretion in place.

To prevent disintegration, bone and ivory must first be dewatered by soaking in ethanol baths for a month with weekly changes. Upon removal, as soon as the surface dries, soak the object in a 10 percent (by volume) solution of polyvinyl acetate (PVA) for about a month. This should stabilize it. A low molecular weight of PEG, about 100, has been used in place of PVA, with some success. A polyurethane coating will help prevent flaking and keep the object together.

Jewelry

The best advice here is to take jewelry to a jeweler. Unless you recover quite a bit of it, the investment in tools and materials will not warrant the expenditure.

If you want, you can purchase a small ultrasonic cleaner for home use. This device runs on household current, and uses sound waves to break free dirt and debris. It consists of a vibrating unit and a tub for the soaking solution. Commercial solvents are available, but my jeweler uses Top Job. For light encrustation, check your jewelry every fifteen minutes. For gross concretions, hours, perhaps days, may be required to completely sound off undesirable marine growth.

Final polishing should only be done with steam units and miniature, high-speed wire brushes and buffers—specialized tools which your jeweler will already have. Marine conservationists cannot be found on every street corner, but jewelers can. So, where you have the opportunity to utilize expert help, that is the way to go. Not only do you get the use of the latest technical equipment, you also get years of expertise as well. If a recovered watch needs replacement parts, he can obtain them. If a gold locket needs repair, he can fix it. If a silver bracelet needs burnishing, he has the proper instruments to do the job right.

For something as delicate and as valuable as jewelry, it is money well spent.

Afterword

Time is of the essence. As you can see, many of the procedures to adequately preserve or restore marine artifacts described herein require months, if not years. But this time is, in a sense, purely subjective—vast amounts of work are not required on your part, only patience.

Although we are all eager to display our finds as soon as possible, this eagerness is self-defeating, as the artifacts must necessarily suffer for this haste. Portholes that were painstakingly polished will have to be repolished or completely redone; deadeyes that were improperly preserved will be lost irretrievably.

Take the time to do it right the first time and you will be rewarded. Then you can display your artifacts with pride, and without fear that they will be destroyed due to your lack of industry.

Any artifact worth taking is worth preserving.

CHAPTER 8

Quick Photography Techniques

Taking pictures in deep, dark, murky waters where bottom time is limited requires procedures different from those the conventional underwater photographer is used to. Many of the methods used in clear, shallow water are either impractical, unwieldy, or take too long to execute in this different environment. This chapter does not seek to supplant the accepted procedures taught in underwater photography courses; rather, it augments those teachings for the specific problems relating to conditions where poor visibility prevents the photographer from seeing his own equipment, where narcosis interferes with the making of necessary adjustments, and where situations such as penetrating an intact shipwreck present extreme difficulties.

Camera System

While double-strobe systems with long extension arms operate well in allowing overall lighting without creating shadows, and do a good job of eliminating backscatter, the dynamic drag of such a setup against a stiff current makes it undesirable. A camera and strobe assembly that is compact is not only easier to carry, but easier to handle and use as well. The fewer control mechanisms that need attention, the better.

A lanyard that is tied tightly to the camera and fitted snugly around the wrist frees both hands for use: you can let go of your camera when you need to pull hand over hand down the anchor line, without fear of losing your valuable equipment. You can alternately take pictures on the bottom, or drop the camera and let it dangle when you need your hands for other chores.

Equipment lanyards are necessary for accessories, even though they are attached by setscrews and knurled knobs. These mechanisms have a habit of working loose and allowing the device to drop off or float away. Lanyards don't look professional, you say? Forget about imagery and come down to practicality. More than once I have returned from a dive with my light meter hanging by a string, or my viewfinder floating like a buoy.

By attaching an underwater light to the strobe bracket, you eliminate the need for a third hand to aim the beam. When the light is aligned properly, it does double duty both as a camera pointer and as your main light source. Some lights are outfitted with detachable handles connected by a quarter/twenty nut and bolt. Remove the handle, use a longer bolt, and secure it directly to the strobe bracket. For lights without this form of attachment, a flat metal or plastic rod can be bolted perpendicularly to the strobe bracket and the light attached to it by stainless steel hose clamps. Now, even deep inside a wreck, you can see while taking pictures without having to drop your camera to pick up your light.

The viewfinder must be adjustable. Having one that is set at a predetermined distance forces you to make parallax corrections in your head. When diving to a depth of 200 feet you are too rushed, and have too many other things on your mind, to worry about whether or not you are half framing your subject. With the viewfinder preset in accordance with your dive plan, you can look through the optical lens and know that what you see is what you are shooting. Also, it is desirable to mask the viewfinder with a set of crosshairs that can be lined up with the rear center dot. This helps to keep you from looking through the lens sideways and seeing an offset view.

One of the two most important items of a quick photography system is a wide-angle lens, the wider the better. This has many applications. It allows you to get extremely close to your subject in order to reduce the amount of particulate matter between it and the lens. In poor visibility a picture can be framed which, using a normal lens, would put you beyond the limit of sight. Also, in deep, dark water, or inside a wreck, when the focus and aperture markings of your lens cannot be seen, adjustments must be made by feel. With practice one can learn this skill, but at narcosis depths one is apt to get confused, with undesirable results: either out-of-focus or over- or underexposed pictures. But take, for example, the Nikonos 15-millimeter lens. When set for two feet at f-16, the depth of field is one foot to infinity. Thus, no adjustments are necessary.

The other most important piece of equipment is a fast, powerful strobe. Recycle time must be rapid, and in this respect a strobe with rechargeable nickel-cadmium batteries is preferred over the slower-charging alkaline batteries. It must have a wide enough beam angle to cover the wide-angle lens in use—and then some, in case of misalignment. It must have three power settings. It should have a modeling light, so you can set up your working distance at the beginning of the dive and from then on only have to align the crosshairs of the viewfinder on the light beam. (The modeling light also doubles as a backup light.) And the mounting bracket must be short, but detachable.

Porpoises play and perform for a group of photographers.

Now that I have described the system, let's see why each one of these choices has been made.

Quick Techniques

The methods I will now describe are not only quick but, just as important, require very little thinking to implement. For this section I am going to assume the use of all the equipment suggested above. I will show what can be done with it and demonstrate why I have chosen the type of equipment and the methods I have.

My three most important rules of photography are: always, always, always take your camera. Take it even if you don't think you are going to need it, for that is when the opportunity for that one unbelievable shot will occur. The next most important rule is: don't be afraid to use film. When you see any photographer's work you are seeing only the tip of the iceberg. Chances are he has thrown out 90 percent of his pictures and kept—and shown—only the best. You should do the same.

If you are spending hundreds or thousands of dollars on a dive trip or a photo-oriented vacation, the cost of the film is the smallest amount of the total trip cost. Remember, the object of your trip is not to see how much money you can save on film, but how many good photos you can bring back with you. That is the difference between a snapshooter and a photographer.

As an extension of this, don't let anyone see your bad stuff. Throw out anything that is questionable, and keep culling your material. Like dusting your house, it is a never-ending task.

Bracketing your shots for proper exposure is extremely important, even with automatic camera systems and through-the-lens (TTL) metering. This is because the marine growth and ships' hulls you are photographing will be a compilation of different shades, different contrasts, and different degrees of brightness. What the automatic system may choose as an overall average reading may not offer the best results. A white, highly reflective anemone growing on a dull, rusty hull plate is a prime example. Here, the photographer's experienced eye will stand in good stead. Otherwise, the three-shot technique is highly recommended. You take one frame an f-stop underexposed, one exact, and one overexposed. Out of this set of three, one will undoubtedly be right. If you shoot only one picture, you'll find that one of the other two would have been the better exposure.

The usually prescribed method of doing this is to change the lens aperture. This is not only slow and cumbersome, requiring that the camera be turned so you can see how to effect the change, but practically impossible in the darkness of deep water, in poor visibility, or when inside a wreck.

A quicker way to control exposure is to switch the power settings of the strobe. If you start with your strobe on half power as the optimum for your prescribed distance and film speed, flip it to full power for your overexposure, then to quarter power for underexposure. Because the switch is on the back of the strobe, that portion which is facing you, it can be adjusted virtually by feel and without much thought. If you are taking a series of shots, instead of just carrying the camera in case you see a good opportunity, take one group of pictures at high, medium, and low, and the next group at low, medium, and high. This way, you can never forget which of the exposures you have taken. Otherwise, leave the strobe set on the optimum setting in case a marine animal gives you time for only one shot.

Another, even simpler, method is to move the entire camera system. Compose the picture with the viewfinder at your face mask, and press the shutter release. Then extend your arms partway and shoot again. Finally, lean forward and lock your elbows and take the last shot. You have just taken three photos, each a foot apart, with approximately one f-stop difference between them—all from the same spot and without any control manipulations whatsoever.

I use the same procedure of progressive photography when swimming toward a possible subject. Again, without touching any controls other than the shutter release, I simply keep snapping during my approach. This makes each successive exposure a little brighter than the last. Now you can see the advantage of a rapidly recycling strobe—you can't afford to twiddle your thumbs for ten or fifteen seconds between shots while your subject swims off, or silt obscures it.

Alternatively, you can detach the strobe and, while keeping the camera in the same location, move the strobe in and out, thus varying the intensity of light that falls on the subject. An advantage of this method is that it allows for creative side lighting, while further eliminating backscatter. This side lighting can be important in casting shadows where the subject matter, such as a porthole, is without relief. The shadow will add a third dimension to the picture and take away the flatness.

Composition bracketing is also important in wreck photography. One picture, for example, may show a ship's wheel from a distance, illustrating where it lies with respect to the wreck, while a close-up shows detail and marine growth. Ideally, since ships' wheels are such a rarity, for a photograph of such importance one should take a minimum of nine pictures: three bracketed exposures from a distance, three from midpoint, and three close up. Remember, how often will you see this sight? And how important is it to have that one great shot?

Move around your subject to capture it from different angles, and bracket each angle. Also, don't forget the vertical format. Some subjects, particularly upright divers, fit the vertical frame better. When in doubt, take several of each.

Shipwreck Techniques

Shipwrecks, whether iron or wood, create their own special hazard to underwater photographers: silt. Rusting iron and decomposing wood look clean until you touch the surface or get close enough to cause a current around it. Then, billowing swirls of fine particulate matter try hard to obscure your subject. In this situation, speed is even more important.

The first precaution you can take is to move slowly, and try not to overuse your flippers. Even so, this will not prevent silt from being kicked up, but will only slow down its appearance.

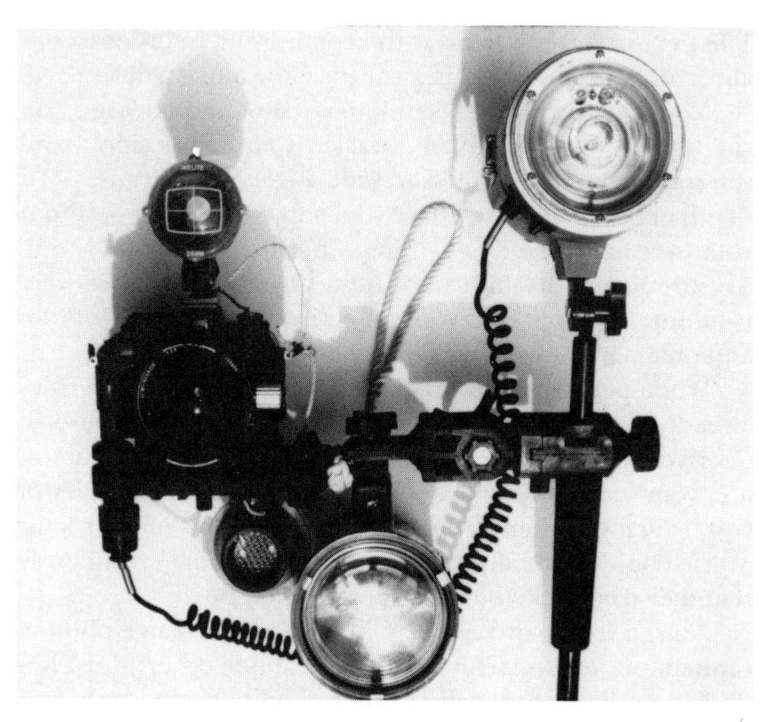

A typical wreck diving camera setup. With the dive light mounted to the bottom of the tray, the photographer can always see what he is going to shoot, especially inside a wreck, or in deep, dark water.

Always have your camera handed to you, as the shock of hitting the water may loosen seals.

The next best thing to do is to keep moving in order to stay ahead of it, as you would in a car traveling along a dusty road.

Now, by applying the techniques of shooting as you go, you can bracket your exposures, bracket your composition, and still stay ahead of the silt. The wide-angle lens, because of its depth of field, will always have your subject in focus—without your ever having to do anything other than swim and shoot. At extreme depths this can be quite an advantage. In a twenty-minute, 200-foot dive you can easily shoot off a roll of film and obtain many worthwhile pictures.

When penetrating a wreck, besides following all the rules already mentioned, you must work even faster because of debris descending from overhead. No matter how careful you are, your exhaust bubbles will rise and dislodge particles of rust, which will then drop off and fall in front of your lens. Most people are totally unaware of this, thinking erroneously that they have kicked up the silt with their fins.

Now, a final word about film. Most underwater photographers prefer Kodachrome 64 and abhor other films. Kodachrome 64 has excellent color rendition, contrast, and balance. But in deep, dark water, where unfortunately many shipwrecks lie, it is not practical. Ektachrome 200 or 400 on the other hand may not be as brilliant in color, or have as much contrast, but a slightly less colorful picture is better than no picture at all, one that is vastly underexposed, or one that is out of focus or blurred as a result of wide aperture settings and long exposure times.

The differences between ISO 200 or 400, other than film speed, are very little; so you may as well go with 400. This is 2½ stops faster than ISO 64, and the extra latitude is essential. For available-light photography, try the Kodak P800 for slides. This film can be pushed to 1600 with acceptable results, although generally speaking 3200 is pushing things a bit too far—taken in conjunction with water motion and microscopic silt, it tends to get grainy.

A friend once asked me, after I returned from a dive trip, whether I had a good time. My reply was, "I'll let you know when my film is developed." When your eye is glued to the back of a camera, you see little of what is actually happening. Don't let anyone tell you differently: photography is work. But when you spend the time and effort to learn the necessary procedures, and act accordingly, the experiences will be permanently preserved and will live long after the memories have faded with the passage of time.

What was but a fleeting moment in your life becomes eternal.

CHAPTER 9

Shipwreck Research

Perhaps the most exciting adjunct to diving wrecks is delving into history.

Ships are built with forethought, planning, and great precision. They ply the oceans anonymously, manned by nameless crews, promoting commerce with cargoes diligently grown, distilled, or manufactured. To and from the far corners of the earth they carry people, cargo, and precious liquids. Without ships, a worldwide civilization would be impossible.

Yet we seldom assign any significance to these freighters, tankers, and luxurious passenger liners until the ultimate calamity of death and destruction brings them to our attention. Then we realize what we have lost, and strive to bring back to life that which has slipped our grasp.

Like a paleontologist studying the bones of prehistoric beasts, wreck divers descend upon the skeletons of sunken vessels, to observe these wood or steel hulls before the ravages of time and the sea dissolve them away forever. We touch a piece of history, and we are in awe.

Word of mouth, usually from fellow divers, sparks the first interest. The facts may not be accurate, but they are exciting and we thirst for more knowledge. As we hear conflicting accounts of the sinking and human suffering, we must satisfy our curiosity and discover the real truth: unbiased, untarnished, and unconfused by the retelling, where pertinent data is jumbled, lost, or made up to fill in the gaps.

A glance at a nautical chart may show where a sinking occurred, but it is to the newspapers we must turn for a contemporary description. Daily periodicals are notorious for obfuscating facts, for sensationalizing, for adding glamor, for inexactitude of reportage. Columns are written quickly, thus inaccurately: their basic intent is to increase circulation, and this must be taken into account. You can get a general feeling for what happened, but distrust the intimate details. Witnesses are often misquoted; mistakes run rampant. I once read a newspaper story about myself, and seriously doubted that I was even there at the same incident, so mixed up was the description.

Bearing in mind that newspapers do not always conform to the gospel truth, the local gazettes can still be a valuable resource, often carrying a fuller story than the nationals because of local interest. You may glean facts otherwise unavailable. The drawback is that you must travel to the shore towns where each paper was printed. In addition, you must know the exact date of the casualty.

Most large city libraries keep the *New York Times* on microfilm. This is a superior newspaper, well written and authenticated, despite the use of wire services over which the paper has no control. Better yet, it is indexed all the way back to 1859. All you need do is pull out the appropriate year, or run an annual check, and look under the headings "Disasters At

Sea"; the various subtitles under "Marine" or "Shipping," or "Accidents." The listings have changed over the years. For really large stories, try the name of the vessel in question. For the Second World War, look under "World War II—Naval Action."

Rarer in the United States, but obtainable on microfilm at some large universities, is the *London Times,* published since 1787 and indexed since 1797.

The *Readers' Guide to Periodical Literature* is issued twice a month and later bound into an annual volume. It is an index of the major popular and academic magazines. Here you can cross-reference articles by topic, title, or author, to locate historical pieces where the author has had the advantage of time and perspective to verify details. If the author did a proper job, you will have an accurate and concise narrative that weeds out the repetitive dialogue of a newspaper byline, while adding a touch of clarity. That is the purpose of a magazine article. Unfortunately, the author's work is sometimes abused by space constraints and awkward editing.

Books have no strict length, so a story can be told with as much detail as available. Even a small library has a selection of tomes on ships, from Roman galleys to roaming galleons. Get acquainted with the public as well as the college atheneums, where books abound.

In addition, the serious investigator must do original research. By that I mean making your own inquiries into official documents stored in maritime museums and government archives. Unless you live close to these great sources, you must conduct your research by mail. This is time-consuming, but the collected data can be more than worthwhile, not just in terms of interest, but also in terms of the usefulness of the information.

For example, a photograph of a vessel before it sank can divulge valuable secrets about the ship's structure that may help you identify a questionable wreck, or enable you to locate specific artifacts. When requesting pictures from museums, be sure to include as much information as possible about the particular wreck you are researching.

The three-inch stern deck gun of the *Kolkhosnik*.

Many ships have the same name. What differentiates one from another are statistics: building date, sinking date, location, tonnage, length, shipping line. You can get the specifications from the *Lloyd's Register,* the annual volumes of which go back to 1869. For older ships try the Lytle-Holdcamper List, *Merchant Steam Vessels of the United States, 1790–1868.* A good overall beginners' book is the *Encyclopedia of American Shipwrecks,* by Bruce Berman. Of course, all this presupposes that you already know the name of the ship. It gets tricky when you have an unknown wreck you want to identify. This means poring through the indexes of book after book, reading locations of lost ships and trying to match them with your wreck. No one ever said research was easy.

If you can hazard a guess as to the probable era of a shipwreck, your best bet is to sift through the *Lloyd's Weekly Shipping Index,* published since 1869, and the similar *New York Maritime Register.* These list marine ship losses all over the world, the only drawback being their lack of availability and the fact that they record casualties, not just ships wholly wrecked. You must wade through accounts which are of little interest to you.

Practically everything you want to know about ship disasters is contained in government files. Unfortunately, the retrieval of that information is a Herculean task. At the National Archives I once asked for three files by the complicated record group numbers. A little later a clerk wheeled a huge, triple-decked cart from the stacks, loaded with boxes of papers. When I asked which were mine, he said all of them were!

Obviously, if you seek information that is packed away like this, you cannot expect to get anything other than a cursory reply by mail. You must camp out in the District of Columbia and spend painstaking hours, days, perhaps weeks, locating just the right piece of information. Begin your research at home by writing to the National Archives for their free forty-page shipwreck bibliography listing books, charts, and annual reports. Many of the books can be borrowed on interlibrary loan.

Tom Packer checks the anchor chain on the *Andrea Doria*.

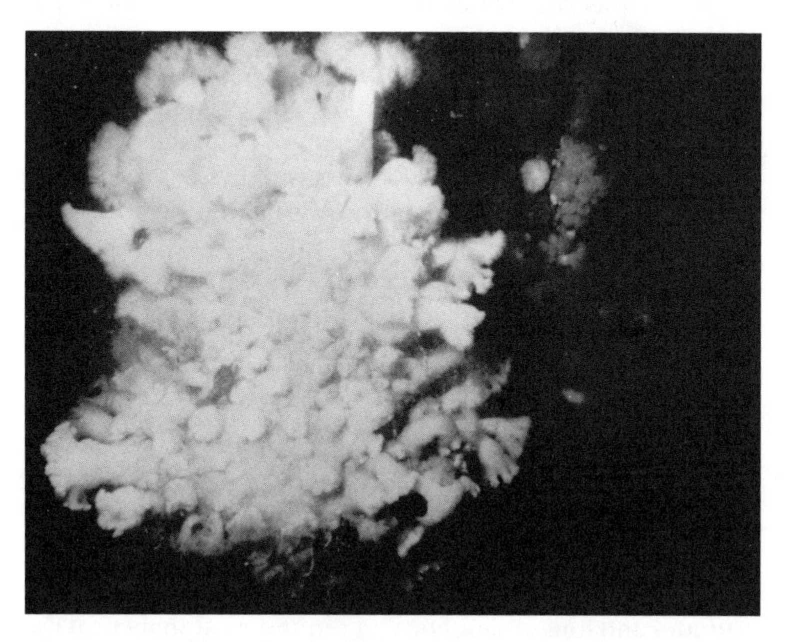

The *Andrea Doria*'s bell found! Under this intense growth of flowering anemones is 150 pounds of silvered bronze.

I won't even begin to tell you about the photo files in the National Archives other than to relate this incident. I arrived one morning with a sandwich in my briefcase, and quickly became engrossed in the tens of thousands of eight-by-ten glossies. When an archivist told me I had only fifteen minutes, I asked him where I could eat my lunch. He explained they were about to close. It was almost five o'clock, and I had completely missed the noon meal!

Down the street is the Library of Congress. In addition to being the largest repository in the country for published books and eighteenth- and nineteenth-century newspapers and marine gazettes, they store the published records of the Life-Saving Service and the Revenue Cutter Service (which merged in 1915 to form the Coast Guard), the Lighthouse Service, and associated material. Their map and nautical chart collection is unparalleled.

As long as you are in Washington, stop in at the Coast Guard office and look through the reams of hand-written Life-Saving Service reports, dating from 1854. For more recent wire drag surveys, the Hydrographic Office can help you out. And in the Washington Navy Yard you will find the Naval Historical Center and the Naval Photographic Center. Contact these federal agencies for more information on what they have to offer.

More recently the National Ocean Service, under the auspices of the National Oceanic and Atmospheric Administration, has computerized its files and issues a printout called the Automated Wreck and Obstruction Information System (AWOIS). The user's guide lists some 200 references, and is itself an incredible bibliography of shipwreck sources.

The AWOIS list, while primarily emphasizing hazards to navigation, covers the entire coast of the United States, including the Great Lakes. You can order the entire file, or buy special areas such as all wrecks and obstructions between specific latitudes and longitudes. You can request an alphabetical listing, or have a listing printed by descending latitude. Information for each wreck includes source and reliability of data, and may offer Loran numbers and historical descriptions. The positions of thousands of unknowns are recorded, just waiting to be matched with lost ships.

This list is growing daily, as input data are received from ongoing official surveys and private sources. When you requisition a printout, your copy contains all the data in the banks up until the precise moment of actual printing.

The appended addresses will give you ample source material to research your favorite shipwreck, locate that elusive position, or discover a previously unknown site. Wreck diving is a stimulating sport with much to offer, intellectually as well as physically. But the collection of data is not an end unto itself, for the information, in order to became part of the great fund of civilization, must be shared.

The fruits of your labors may coalesce in the form of lectures, articles, or just in conversation. Whatever framework you use, disseminating knowledge can be more rewarding than simply squirreling it away in your personal files. And what you offer to someone else will very likely be returned with interest.

Government Sources

National Archives
National Archives and
Records Administration
General Services
Administration
Pennsylvania Avenue and
8th Street, NW
Washington, DC 20408

Hydrographic Surveys
Branch
National Ocean Service
NOAA
Rockville, MD 20852

United States Coast Guard
2100 2d Street, SW
Washington, DC 20593

Naval Historical Center
Washington Navy Yard
Washington, DC 20374

Smithsonian Institution
National Museum of
History and Technology
Washington, DC 20560

Library of Congress
Washington, DC 20540

Maritime Museums

Hart Nautical Museum
Room 5-329, MIT
Cambridge, MA 02139

Maine Maritime Museum
963 Washington Street
Bath, ME 04530

Mariners Museum
Museum Drive
Newport News, VA 23606

Mystic Seaport Museum
Greenmanville Avenue
Mystic, CT 06355

National Maritime
Museum, San Francisco
Bldg. 201, Ft. Mason
San Francisco, CA 94123

Peabody Museum of Salem
East India Square
Salem, MA 01970

Penobscot Marine Museum
Church Street
Searsport, ME 04974

Philadelphia Maritime
Museum
321 Chestnut Street
Philadelphia, PA 19106

South Street Seaport
Museum
16 Fulton Street
New York, NY 10038

Steamship Historical
Society Collection
University of Baltimore
Library
1420 Maryland Avenue
Baltimore, MD 21201

CHAPTER 10

Lobster: Tail or Tale

While most divers drool over the prospect of eating lobster tail sauteed in butter, there are many who return from a hard day's diving with nothing to show for their efforts other than the story of the one that got away. And while these tales are usually embellished with drama, pathos, and intrepidity, they do not, in a manner of speaking, bring home the lobster tail. Hungry friends and an expectant family want more than excuses to top the seasoned rice on their dinner plates. So, since everyone is already expert in talking about lobsters, this chapter will tell you the real story: how to catch them.

Lobsters, believe it or not, are not ferocious monsters. In catching them you don't lay your life on the line by engaging in mortal hand-to-hand combat. Neither do you need to be a hulking brute in order to wrestle a lobster out of its home. You need only two things: tools of the trade and knowledge of the enemy.

It has often been said that a good eye and a fast hand, aside from the obvious scuba gear and safety equipment, will get you all the lobster you can eat. While this is basically true, there are a few items that may increase the likelihood of your success. I speak of course of the underwater light which will not only highlight the beast's reddish shell but will prevent you from banging your face mask in the sand on those days when visibility is not the best; the mesh bag in which to store your catch (hugging a lobster to your chest or clipping it onto your buddy's loose beaver tail can have unsatisfactory results); and the "bug hook."

There is a strong dichotomy of opinion about the use of the lobster gaff. Proponents hail it as the greatest invention since the lobster pot; detractors deride it as unsporting. But when you get up at five o'clock in the morning and drive a hundred miles to the shore and pay exorbitant boat fees to spend the day lumping around the ocean suffering more than a touch of nausea, being a good sport doesn't put food on the table. And who said this was a sport, anyway? Some may even misquote, "It isn't how you play the game, it's whether you win or lose." Moral issues aside, I'll return to bug hooks later and tell you not only how to make them, but how to use them—humanely.

Although lobsters roam the ocean floor freely, they are too scarce to locate over the vast sandy tract with the limited bottom time available. However, they tend to congregate in such places as rock piles and jetties, where they can back their unprotected hind quarters into a safe retreat, or where they can lie concealed under a cover of kelp. Undoubtedly the highest concentrations of lobsters are found in the innumerable cracks and crannies offered by sunken ships and collapsing hull plates.

Lobsters are night foragers. Shy by day, during the nocturnal hours they roam the ocean bottom seeking out tasty morsels such as mussels, crabs, and starfish, or fish that are either slow or dead. They may dine out, or drag their prey back to the security of their humble abodes. Obviously, the best time to catch them is at night when they are out in the open. During the day, however, you will find them silently sequestered between the beams of rotting sailing ships or under the steel plates of rusting freighters.

The first giveaway of an occupied hole is the flaunted claw. This is done to intimidate possible usurpers, such as rock crabs and ocean pouts, who also like to live in tight-fitting cavities. The claw is equivalent to a "no vacancy" sign to the local denizens. To the diver looking for fast food, it's the same as the golden arches.

Less obvious but distinctly visible to the keen-eyed diver are the protruding antennae, by which the lobster senses invaders. This means that the lobster is alert, feeling with this appendage for what it cannot see. You must be careful not to brush up against the antennae or even to cause mild currents around them, for the lobster will instantly become aware of your presence and retreat deep into its cave.

No one, not even a lobster, can spend a whole day in a dark hole without feeling the call of nature. As it sits in happy seclusion, calmly digesting last night's dinner, it must perform other bodily functions as well. Lobsters, like people, enjoy a clean nest, so you can well imagine that the piling up of excretions can become annoying. Conscientious housekeeper that it is, the lobster uses its swimmerets (two rows of appendages under its tail) to sweep out debris in the same way you would use a broom—except that it has four pairs of them! Since the lobster doesn't use a dustpan, and doesn't subscribe to a garbage collection plan, it merely brushes its refuse out the door.

The astute diver will keep a sharp eye for sand turned dark by the admixture of excrement. Sometimes, too, you will find that the lobster stores its leftovers nearby in case it decides on a midday snack, so look for meat scraps and crab shells.

Even an apparently empty hole may still conceal a clever lobster. In a large cavity you must inspect for side passages. In a small opening you might have to dig out sand which the lobster has built up as a barrier—an effective protection against most marauders. Some holes have more than one entrance, so that you can scare a lobster away from one and catch it at the other. Inside wrecks, don't forget to look up, for sometimes a lobster will inflate its air bladder and float along the bulkheads or settle in an upper corner. The lobster hunt is truly three-dimensional.

I once found a lobster in the middle of a ten-foot pipe, out of reach from both directions. I tried taunting it out by sneaking up on its back side and touching it with my crowbar. (Every self-respecting wreck diver carries a crowbar.) After several futile attempts I finally swung the crowbar like a clapper, ringing the inside of the pipe like Quasimodo clanging the church bell. In seconds a dazed and disoriented lobster shot out the other end to cower on the open sand where I picked it up.

Assuming that you've been able to locate a lobster, we now come to the most important segment of the tale: the grab. Ideally, you should sneak up on the lobster from behind, where it is relatively defenseless, and grasp the tail firmly just behind the thorax. Don't relax your grip or it will either squirm away or shoot off with rapid-fire scoops of its powerful tail muscles. Watch out for the smaller guys, for they have an amazing backhand reach and will think nothing of nipping your forearm if you're not careful.

Unfortunately, lobsters don't usually present their tails for such overt mishandling. When you look into a hole you're likely to be facing the business end of their very formidable armament: two huge claws which can rend, tear, pinch, and cut delicate human tissue and which can have deleterious consequences on extended digits.

Notwithstanding the possible detrimental side effects, the key word is audacity. The timid diver rarely gets his meal. You must act quickly and decisively, for he who hesitates goes home empty-handed—or worse, with aching fingers and wounded pride.

Again, as a personal aside, there are exceptions to this rule. Once I laid my goody bag on the bottom and opened it up to take out a hammer and driftpin punch (other tools carried by every self-respecting wreck diver) and left it in order to work on a porthole overhead. After many minutes of furious hammering I looked back to find that an irate lobster had come out of a nearby hole to see what the disturbance was about. By chance it was standing within the metal perimeter of the hoop, so I just reached down and snapped it shut. The lobster was even more unhappy when I later threw my tools and a porthole on top of it and gave it a liftbag ride to the surface. But let's get back to the more normal procedures.

The most direct method is to reach into the hole with one hand and seize the adjacent claw. Then, when the lobster tries to crush your hand with the other pincer you grab that one too. If you cling to the claws adamantly, the lobster is effectively harmless. But that doesn't put it in the bag, and many divers who have gotten this far still lose their prey.

Big lobsters have a lot invested in their claws and are reluctant to give them up. But a small lobster will make the sacrifice in the blink of an eyestalk, and leave you holding the claw. You may have quite a tale to tell, but not much edible for your effort.

In order to prevent the lobster from dropping its claws and retreating into the inner sanctum of its hole (where it will eventually regenerate any lost parts) you must not grab the meat end but take hold at the first joint where the appendage is attached to the body. The way the lobster releases the claw is by relaxing the muscle of that first joint and scooping backward with its tail while you are pulling forward. But by holding the joint to take away the leverage the lobster would have when you hold the pincer, you don't allow it to flex those two intervening knuckles. It's even better if you can also take hold of the first walking leg, for then its position is untenable.

More common, perhaps, is the one-hand grab. In this instance you grasp the lobster from only one side, again trying to reach far enough back so you can gather in the first walking leg. If you keep your arm low, the mere presence of the claw you are holding prevents the other claw from getting you, for the dull-witted crustacean will never think to lift one claw so it can get you with the other. Note: if grabbing a one-clawed lobster make sure you grab the side that has the claw.

At this point you can still reach an impasse, and get only the crust of the crustacean. Once a lobster realizes that it can't pinch you, and feels itself being pulled out of its hole, it initiates a defensive plan. It will instinctively push its claws down (which is why it can't get you), which action pushes its head against the ceiling of its domain, thus forming an almost impregnable wedge. The horn which projects past the carapace and which protects the compound eyes and their stalks from harm terminates in an upturned point and has running along its length protuberances all of which act as brakes.

Shake well before opening is the motto. By vigorously jerking the recalcitrant lobster back and forth, you can loosen its grip and move it forward by degrees. Another method is to exert strong sideways pressure against the body so that the lobster is jammed against the side of the hole. Once it is immobile, you can work your fingers along the carapace until you reach the tail, then swing the lobster around and drag it out backwards, making its wedge ineffective.

Again, I want to reemphasize boldness. The only way to get ahead in the world is by taking the beast by the horns. Even if you can't see the lobster you must plunge your hand into the hole, determining the lobster's position from its antennae, perhaps, or relying on the shock value of your sudden intrusion. This may sound rash to the uninitiated, but it usually produces results.

The general rule of thumb in catching lobster is: if it works, don't knock it. Once you've spotted one, turn your light away immediately as the beam will tend to scare it off. On the other hand, if a lobster is too far back in a hole, or if you made a grab and missed and it has receded beyond reach, try flashing your light rapidly in front of the opening: this action may arouse curiosity.

Mum is the word, for usually lobsters are alerted by noise and will be on the alert. However, noise can be used to attract a lobster from the depths of its cave. A light tapping with your dive knife around the opening can whet its interest. A loud banging in an area behind the lobster can scare it out or cause it to turn around.

Lobsters are territorial. They will protect their turf against invading fish, inquisitive crabs, and other itinerant lobsters.

The classic lobster position in defense of its hole: antennae out, and claws on guard.

Large lobsters, like this thirteen-pounder, are best handled by grasping the tail. The claws cannot reach back to grab the grabber.

Try this gambit. Place a starfish or a rock crab, both ubiquitous wreck inhabitants, in the entrance of a lobster hole. Continue your dive and circle back a few minutes later. You may find your elusive lobster either feeding on or in the process of ejecting the intruder.

A friend once did something similar to this with less than desirable results. Knowing lobsters' habits, he took a small lobster out of his goody bag and forced it into a hole already occupied by a larger relative, who was wise as well as aged. The small fry, thinking to make good its escape, charged into the hole. The hole then clouded up as furious swimmerets dug up the silt and masked its retreat. A few seconds later, out of the artificial darkness, the little lobster was tossed—a piece at a time!

If all else fails, there's always the gaff. The gaff is a simple contrivance either store-bought or homemade. The commercial variety is actually intended for use in landing large game fish. It is two to four feet long, made of tubular aluminum with a plastic hand grip, and presents a hook that would fit around an orange.

But you can make a better one with an old fishing pole, a two-aught hook, and some strong twine. The rod can be five or six feet long and made of solid plastic or fiberglass, plastic being the better material since it doesn't splinter when pinched by a pugnacious lobster. If you use one with a cork handle, remove the handle, as it may come loose at a most inopportune time.

Modify the rod by cutting off the guides. With a jeweler's file make a narrow groove at the tip, the width of the hook shaft. Place the shaft of the hook in the groove so that only the curvature protrudes beyond the end of the pole. With the twine knotted through the eye, carefully wrap it around the rod, keeping it taut and closely spaced, until you reach the curve of the hook. Then wrap it around back down the shaft and tie it off at the eye. A piece of electrical tape will protect the twine from abrasion.

Now for the most important step for any gaff: the point must be filed off bluntly and, in the case of the homemade job, the barb must be crimped or filed flat. The purpose of the bug hook is not to spear or injure the quarry, but to entice it out of its hole. My first claim is humaneness, but I can also apply reason.

The soft underparts of the lobster are extremely vulnerable to puncture. Any damage done, whether or not you eventually get the lobster, inevitably results in death. If the lobster escapes but later dies, you've wasted it. If it's a female laden with eggs, you've killed an entire generation. If you tear it apart you can (and should) suffer the abuse of your fellow divers. And if that's not enough for you, the meat will probably spoil before it reaches the cooking pot. The gaff is a tool, not a weapon.

Use the bug hook delicately. With it you can reach way into the back of a hole and quite often, simply by passing it over the lobster's head and touching it on the tail, induce the lobster to walk right into your arms. If this doesn't work, try grappling the hook around the walking legs and applying a twisting motion so you don't lose the lobster, then pulling gently forward until you can reach it with your hand. Don't get rough or the legs will snap off. And just keep playing until the best one wins.

A further caution concerning gaffs: they can also be dangerous to other divers. You must exercise discretion at all times during your dive. To play it safe, have your gaff handed to you after you've left the boat, and hand it up before you climb up the ladder. Don't poke anyone while swimming, and be extremely careful on the decompression line when there are other divers going down, coming up, or just hanging around.

Be careful of other accidents on the anchor line. Make sure your goody bags are snapped shut—it's a shame to lose what you've worked so hard to get. At the price of lobster on the open market you want to make sure you get what you paid for. Carry two goody bags, one for big brutes and one for small fry—during the long haul to the surface, many enraged lobsters will think nothing of misplacing their aggression on their smaller brothers. It's quite easy to return from a successful dive with a bag full of limp and broken bodies and one victorious monster.

One of the most ignominious feelings occurs when a lobster from your own bag, which may be clipped to your weight belt or dangling from your wrist, gives you a goose through the mesh. It can even be worse when it grabs your buddy and he smites you with his crowbar.

Some last words for ecology: if they're short, put them back. The legal limit for a lobster is 3 1/8 inches from the eye socket to the end of the carapace. Everything looks larger underwater, and nobody carries a gauge on a dive, but I usually wrap my hand around a suspect before I throw it into the bag. If I can see carapace shell on either side of my glove, it's a keeper. Remember, you can always get it next year when it's grown up.

Enough for the cradle robbers, now for the matricides. Lobsters with eggs cannot be taken for any reason, under any circumstances, no matter how large a record breaker. Sometimes it grieves me to have to put back a big momma, but it must be done. Always check for a clutch of eggs by inspecting the lower abdomen *before* placing a lobster in the bag. If you make a mistake you endanger the expectant mother and the future generation she is carrying by putting her through the rigors of capture, not the least of which is possible embolism. You may also engender harassment from your fellow divers on the boat when they discover your lack of observation.

If you do make a mistake and bring a bearing female on the boat, it's your responsibility to put her back. And by that I don't mean climbing to the flying bridge and seeing how far you can fling the hapless creature. If you're making another dive, or someone else is going down, take the lobster back to the bottom and place her tail-first in front of a hole. If not, lean over the rail and gently submerge her in the water. Then she has a fairly good chance of reaching the sand and finding another home before falling prey to a natural predator.

Following all the rules, you can still get a meal fit for a king—at the price of a king's ransom. But then, how exciting is a visit to the fish market?